P9-DXH-669

Oil Spills and the Marine Environment

A Report to the Energy Policy Project of the Ford Foundation

Oil Spills and the Marine Environment

Donald F. Boesch
Carl H. Hershner
Jerome H. Milgram

Ballinger Publishing Company • Cambridge, Mass.
A Subsidiary of J.B. Lippincott Company

Published in the United States of America by Ballinger Publishing Company
Cambridge, Mass.

 First Printing 1974.

International Standard Book Number: 0–88410–312–2 H.B.
0–88410–326–9 P.B.

Library of Congress Catalog Card Number: 74-7128

Printed in the United States of America

Library of Congress Cataloging in Publication Data

Boesch, Donald F.
Oil spills and the marine environment.

"Reports to the Ford Foundation's Energy Policy Project."
Includes bibliographies.
1. Oil pollution of rivers, harbors, etc. 2. Marine ecology. I. Hershner, Carl H., joint author. II. Milgram, Jerome H., joint author. III. Ford Foundation. Energy Policy Project. IV. Title.
QH91.8.04B57 363.6 74-7128
ISBN 0–88410–310–2
ISBN 0–88410–326–9 (pbk.)

Contents

List of Table and Figures

Table

Figures

Foreword

In December 1971 the Trustees of the Ford Foundation authorized the organization of the Energy Policy Project. In subsequent decisions the Trustees have approved supporting appropriations to a total of $4 million, which is being spent over a three-year period for a series of studies and reports by responsible authorities in a wide range of fields. The Project Director is S. David Freeman, and the Project has had the continuing advice of a distinguished Advisory Board chaired by Gilbert White.

This analysis of "Oil Spills and the Marine Environment" is one of the outputs of the Project. As Mr. Freeman explains in his Preface, neither the Foundation nor the Project presumes to judge the specific conclusions and recommendations of the authors who prepared this volume. We do commend this report to the public as a serious and responsible analysis which has been subjected to review by a number of qualified readers.

This study, like many others in the Project, deals with a sensitive and difficult question of public policy. Not all of it is easy reading, and not all those we have consulted have agreed with all of it. Nor does it exhaust a subject which is complex, controversial, and importantly obscured by major gaps in the available scientific data. The matters it addresses are of great and legitimate interest not only to those who are investing heavily in refineries and other petroleum producing and shipping facilities, but also to those who live and work in the areas potentially affected by oil pollution, and in one way or another to citizens throughout the country; the perspectives of these interested parties are not likely to be identical.

In this last respect the present study reflects tensions which are intrinsic to the whole of the Energy Policy Project—tensions between one set of objectives and another. As the worldwide energy crisis has become evident to us all, we have had many graphic illustrations of such tensions, and there are more ahead. This is what usually happens when a society faces hard choices, all of them carrying costs that are both human and material.

But it is important to understand that there is a fundamental difference between present tension and permanent conflict. The thesis accepted by our Board of Trustees when it authorized the Energy Policy Project was that the very existence of tension, along with the inescapable necessity for hard choices, argued in favor of studies which would be, as far as possible, fair, responsible, carefully reviewed, and dedicated only to the public interest. We do not suppose that we can evoke universal and instantaneous agreement, and still less do we presume that this project can find all the answers. We do believe that it can make a useful contribution to a reasonable and democratic resolution of these great public questions, one which will serve the general interest of all.

The current study is a clear example of what we aim at. Its discussion of the causes, effects, and prevention of marine oil pollution draws upon a complex and inter-related body of knowledge comprised of marine biology, oceanography, biochemistry, physics, zoology together with information about oil drilling and transport practices, fisheries and wildlife preservation, government regulation, and details of specific episodes of oil pollution. The study identifies the base of scientific and technical knowledge upon which public policy makers must rely and it offers an agenda for research to improve that base. The authors have treated a hard subject with the respect it deserves, and I commend their analysis to the attention of the American public.

McGeorge Bundy
President
The Ford Foundation

Preface

The Energy Policy Project was initiated by the Ford Foundation in 1971 to explore alternative national energy policies. The papers in this book are part of a series of studies commissioned by the Project, and they are presented here as a timely and carefully prepared contribution to today's public discussion about oil spills. It is our hope that each of these special reports will stimulate further thinking and questioning in the specific areas that it addresses. In the interest of timeliness, we are releasing the special reports as they are completed rather than delaying their release until the final report of the Energy Policy Project. However, each special report deals with only a part of the energy puzzle; our final report, to be published later in 1974, will attempt to integrate these parts into a comprehensible whole, setting forth the nation's energy policy options as we see them.

The papers in this book, like the others in the series, have been reviewed by scholars and experts in the field not otherwise associated with the Project in order to be sure that differing points of view were considered. With each book in the series, we offer reviewers the opportunity of having their comments published in an appendix, but in this case no reviewer chose to do so.

The papers in *Oil Spills And The Marine Environment* are the authors' reports to the Ford Foundation's Energy Policy Project, and neither the Foundation, its Energy Policy Project, nor the Project's Advisory Board necessarily endorse their contents or conclusions.

S. David Freeman
Director
Energy Policy Project

Introduction

In the first part of this volume Dr. Boesch and Mr. Herschner tell us that scientific knowledge about oil pollution in the marine environment is at a very primitive stage. The long-term effects of chronic, low-level contamination by hydrocarbons are largely unknown. Even the short-term effects of massive, widely scrutinized accidents remain highly controversial. Decisions made in the next few years to expand offshore oil production and related facilities will necessarily be based on very limited ecological information.

Dr. Milgram, in discussing technological aspects of the oil spill problem, reminds us that oil spills result from human as well as technical error. Thus, solutions must address the institutional side of oil-handling procedures, including regulations, contingency plans, and personnel training.

The Energy Policy Project commissioned these two papers—one on the ecological, and the other on the technological aspects of oil pollution—to bring together in one place an overview of current knowledge about the effects of oil spills and the efficacy of preventive safeguards.

The United States will continue to depend largely on oil—at least for the remainder of this century—to fuel our transportation system. The nation will also continue to use oil as in industrial energy source and increasingly as a raw material for important products such as plastics and fertilizers. Whether produced in and around the United States or imported from abroad, oil will continue to pose risks of polluting accidents.

As plans progress for expanding our supplies of petroleum through domestic production in northern Alaska and in undersea fields on the nation's continental shelves, the risks of oil pollution are being debated in the press and other forums. The prospects for rapid construction of refineries and other petroleum facilities put oil spill discussions on the agendas of federal, state, and local agencies responsible for approving or disapproving such facilities. Growing international trade in oil is also making spill prevention and control the subject of international negotiations and treaties.

Oil spills involve ships, pipelines, storage tanks, fuel trucks, producing wells, and unknown sources. Spills stem from human negligence, "freak" accidents, inclement weather, and unreliable hardware. Impacts occur in oceans, rivers, marshes, beaches, and inland. People, as well as birds, fishes, and invertebrate organisms are affected.

A few examples illustrate the continuing seriousness of this problem. In Milford Haven, Wales, supertankers of up to half a million deadweight tons load and unload oil cargoes daily, and industry officials take pride in Milford Haven's having the world's most modern safeguards to prevent oil pollution. Yet in August 1973, the small tanker *Dona Marika* ran aground at Milford Haven in a storm and spilled 20 to 25 thousand barrels of oil over a period of several days.

The United States' worst tanker spill of 1972 was in Portland, Maine, where the Norwegian ship *Tamano*, under charter to Texaco, hit "Soldier's Ledge," a well-marked rock inside the harbor. Because of a jurisdictional dispute, industry and government officials delayed several hours before taking action to pump the oil from the ruptured ship and begin cleanup operations. A good part of the 2,500 barrels spilled was never recovered. Cleanup procedures continued for over a month, prolonged by difficulty in finding a dumping ground for oil-soaked debris.

The special problems of oil spills in fast-moving rivers were demonstrated in January 1973 when four barges under tow and carrying diesel oil collided with a highway bridge (and each other) in the Mississippi River. Almost 12 thousand barrels of oil was spilled, and by the next day the slick covered 160 miles of the river below the accident site in Helena, Arkansas. Containment of the oil proved impossible.

The technological and ecological aspects of oil pollution invite—indeed demand—further research. Drs. Boesch and Milgram suggest that several areas in their fields are in particular need of careful scientific investigation. In Dr. Boesch's view, the fact that oil-related biological research has been most generously funded by the oil industry itself is evidence of under-emphasis on this problem in the public sector. The fruits of oil industry research, he points out, are often used as evidence in litigation without being subjected to review by the general scientific community. In view of the high level of controversy in the field, it is essential to safeguard the objectivity of scientific data by producing as much information as possible in the public sector. Having accomplished that objective, government agencies will then have the task of incorporating research results into their decisions about future oil policy.

Continuing efforts are also required to protect the nation's citizens against personal or collective losses resulting from oil spills. Traditional remedies of liability are available, but do not cover all the risks. New laws are being sought in this area. For example, the state of Florida recently passed an oil spill law—and successfully defended it against challenges extending to the Supreme Court—that imposes unlimited liability on shippers responsible for oil spills within the state's waters. But new insurance arrangements are also needed to make the absolute liability laws workable.

Regulation of oil transport is complicated by its international nature. Not only does the world's fleet of oil tankers move through the waters of all continents and most nations, but individual vessels often operate by arrangements involving several nations. A single tanker may be registered under the flag of one nation while being chartered to an oil company with headquarters in another, being manned by a crew from still another, and carrying oil between yet two others. Safety and operating standards vary widely among the world's nations. The negotiation and ratification of binding treaties to govern oil transport practices should be a priority concern for the United States and other nations. This year's conference on the Law of the Sea, sponsored by the United Nations, will have important implications for oil handling.

The United States Maritime Administration currently offers subsidies to shipbuilders in order to expand our domestic fleet. Oil tankers are receiving a substantial share of these subsidies. While there is controversy about the economic merits of the subsidy program, it does offer an opportunity to impose stringent safeguards against polluting accidents. The subsidies are tied to requirements that new ships meet design standards set by the government. Continuing assessment of design criteria is needed to ensure that the best technology is used in the construction of new ships.

The Energy Policy Project is grateful to the scientists who contributed their efforts to this volume. The contributors share our belief that the oil spill issue is so important and so closely tied to current policy debates that the current state of knowledge needs to be presented to the public. They share our hope that these papers will contribute to an understanding of this critical part of the energy-environment dilemma.

Pamela Baldwin

Part One

The Ecological Effects of Oil Pollution in the Marine Environment

Chapter One

Introduction

As a consequence of recovering, transporting, and using oil, petroleum hydrocarbons are being introduced into the biosphere—and primarily into the oceans—to an unprecedented extent. Estimates of the amount of oil reaching the sea range from one to ten million metric tons per year (1,2,3), with the most probable rate being near the middle of this range. Most of this comes in small but continuous doses from tanker operation, industrial discharge, and on-shore waste disposal practices. Producing less effluent but disproportionately more public attention and research effort are the large accidental oil spills. Prospects for the immediate future are that both types of discharge will increase as transport tonnage and off-shore exploitation efforts increase.

Faced with this problem and the need to make immediate and intelligent decisions affecting our future, government, industry, the scientific community, and the public are dealing with questions for which there are often no conclusive answers. The extreme diversity of opinion, compounded by overblown statements from one extreme or the other, is the result of too little information and frequent misunderstanding of available information. The literature on the ecological effects of oil pollution is surprisingly voluminous, and several extensive reviews of the subject have recently appeared (3,4,5).

Our overview of the ecological effects of oil pollution attempts to answer the need for an evaluation of available information and of the conclusions drawn from this information. Directed to decision-makers, it is intended as a supplement to more detailed reviews and to the scientific literature itself rather than as a substitute for them. Included are a review of available research on biological effects, ordered by biological community type, a summary of information concerning long-term effects, a discussion of methods and how they influence results and conclusions, and an appraisal of the present direction of research dealing with oil.

OIL POLLUTION RESEARCH

Much early pollution was by fuel oils from the bilges and fuel tanks of ships. Legislation passed in the 1920s prohibited the discharge of oily water in coastal waters in Europe and the United States and reflected a growing concern about oil in the sea. After World War II, energy demands burgeoned, and the oil industry rushed to keep pace. Enormous quantities of crude oil began to be transported over the seas, especially from the Middle East to Europe. Recovery commenced of the oil resources of the continental shelf. These developments produced a quantum leap in both the amount of oil being added to the sea and in the potential for massive spills.

Although there were some investigations of the effect of oil on aquatic life early in this century, scientific interest has greatly increased since the 1940s. In the early 1950s much work was devoted to determining the effects of the developing Gulf Coast oil fields on the shellfish and fishing industries (6,7,8,9). In the 1950s and 1960s oil pollution research received an added boost as a number of major oil spills were studied. These include the *Tampico Maru* wreck in Baja California (10), the *Chryssi P. Goulandris* and others in Milford Haven in the United Kingdom (11), the well known *Torrey Canyon* in the English Channel (12), the Santa Barbara "Platform A" blowout (13), and the small but harmful oil spill from the barge *Florida* at West Falmouth, Massachusetts (14).

With the surge in oil pollution research, the studies began to fall into several general categories. The first is that of post-accident studies. These efforts usually suffer from a lack of planning and too little background information to use for comparison. The two most thoroughly studied spills, the *Torrey Canyon* and the West Falmouth spill, received concerted scientific attention because they happened literally at the "back door" of two of the world's most renowned marine research institutes, the Marine Biological Association of the U.K. and the Woods Hole Oceanographic Institution, respectively. Unfortunately, very few stretches of the world's coastline are within close reach of the wide variety of scientists and equipment required to study a spill properly.

The second approach to the study of the biological effects of oil is experimental; selected species are exposed, under laboratory or controlled field conditions, to oil or to the chemical dispersants used for cleaning spills. Such a "bioassay" approach is traditional in pollution ecology. Acute lethal effects of a pollutant are measured over a standard time period (often forty-eight or ninety-six hours). Longer-term tests may also be run using chronic sublethal doses of the pollutant. Recent work has placed more emphasis on understanding mechanisms of chronic sublethal stresses, which may affect physiology, reproduction, and behavior. These are generally felt to be very important but are more difficult to study than lethal effects.

Just as there are limitations to after-the-fact studies of spills, there are also limitations to the bioassay approach. The most basic limitation is that the experiments, conducted under controlled conditions, are not easily extrapolated to responses in natural systems. This is particularly true of oil research that, because of the physical properties of oil, frequently cannot adequately simulate the conditions of exposure to oil experienced in nature. Laboratory bioassays necessarily concentrate on a limited subset of the species potentially affected, and often the experimental species chosen are the hardiest, because these are easier to maintain in the laboratory. The effects on rare or less tolerant species most often remain unstudied. The limitations of the laboratory bioassay approach is that the results have little relevance to what happens in the marine environment and are instructive only in reflecting the relative susceptibility of different species or the relative toxicity of different oils or dispersants.

A closer approximation of the effects of oil spills may be found in field experiments, which have thus far been limited to intertidal areas or artificial ponds and enclosures. However, even here all physical parameters cannot be faithfully reproduced. Legal problems may often attend experimental release of oil into the environment and have impeded some recent research efforts.

Several general comments can be made concerning the methods used in the research available for review. Most research on ecological effects of oil has been more or less observational; that is, it has involved simple techniques to survey the composition and abundance of the biota compared with background or control situations. As Neushul (15) points out, such studies are often capable only of documenting gross or catastrophic effects. The study of the persistence of oil in the biosphere must necessarily involve considerable chemical analysis to trace its long-term effects on marine ecosystems. Because most ecologists lack the necessary expertise in the complex chromatographic and spectroscopic techniques necessary for chemical analysis of petroleum, most ecological studies have not been supported by adequate data on chemical contamination. The value of ecological studies which do not consider persistent contamination has been questioned by Blumer (2). His view is not entirely uncontested, but in light of the growing documentation of the uptake of petroleum compounds by marine organisms and the persistence of oil in the environment, it appears that analyses of the levels and qualitative composition of such petroleum contamination should be an essential part of investigations in the future.

Chapter Two

Properties of Oil

PHYSICAL AND CHEMICAL CHARACTERISTICS

The physical behavior and biological effects of an oil spill seem to be influenced to a large extent by the type of oil spilled. Of concern are both crude oils and their distillate products. The crudes are extremely complex mixtures of hydrocarbons (compounds made up entirely of carbon and hydrogen) with small amounts of sulfur, oxygen, and nitrogen compounds and trace amounts of inorganic and organometallic compounds.

The type of oil, with its attendant physical characteristics, largely determines the thickness and spreading of a slick, the formation of emulsions (either oil in water or water in oil), the attenuation of light and oxygen in underlying substrates, and the effectiveness of various cleanup techniques. The extent of biological damage is alśo apparently related to the type of oil. Obviously such things as the capacity to smother or physically remove organisms hinges largely on an oil's physical characteristics. The toxicity of oil, on the other hand, seems to be a function of its chemical composition.

There are three basic classes of hydrocarbon compounds in petroleum oils: alkanes (also referred to as paraffins or saturates), alkenes (olefins), and aromatics. Alkanes are chains of carbon atoms with attached hydrogen atoms and may be simple straight chains (normal), branched chains, or simple rings (naphthetic). As with most classes of organic compounds the higher the number of carbon atoms in a molecule, the higher its boiling point and the less volatile it is. Low boiling alkanes produce anaesthesia and narcosis at low concentrations and at high concentrations can cause cell damage and death among a wide variety of lower invertebrates (16). Higher boiling alkanes are naturally produced by life processes and are found in all marine organisms. Higher boiling alkanes of petroleum origin are not normally toxic; however, they may affect chemical communication and interfere with metabolic processes.

Alkenes fall between alkanes and aromatics in structure and properties. They are not found in crude oils, but in some refined products, such as gasoline, and in cracking products. Alkenes are probably more toxic than alkanes, but less so than aromatics.

Aromatic hydrocarbons are characterized by the possession of six member rings of carbon atoms which have three double carbon-carbon bonds. These are the "benzene-rings" familiar to basic chemistry students. Compared to alkanes, aromatics are not common in nature. Low boiling aromatics are thought to be the most immediately toxic compounds found in oil. Found in virtually all oils, low boiling aromatics are quite soluble in water and can kill not only at full strength, but in dilutions. Higher boiling aromatics, especially multi-ring compounds, are suspected as long-term poisons and some are known carcinogens (17).

The biological impact of an oil spill depends in part on the type and amount of oil spilled, and the amount of change the oil has undergone while in or on the sea. For example, light fuel oils, having more aromatics than heavy fuel oil, will have more toxic effects. On the other hand, heavy fuel oil might do more mechanical damage to intertidal life by smothering or physically removing organisms. Indeed, these differences have been used to explain the disparity in observed effects of oil spills. For example, Straughan (18) has suggested that the devastating effects of the *Tampico Maru* and West Falmouth spills were due to the fact that light fuel oils with high concentrations of volatile aromatics were spilled. She contends that crude oil spills like that at Santa Barbara are not as devastating because of the lower toxicity of the oil reaching shore.

WEATHERING

Apart from the physical and chemical characteristics of an oil, other factors condition the effects of spilled oil on an ecosystem. These include the degree of change an oil undergoes as it is "weathered" in the environment. Weathering processes include oxidation, evaporation, dissolution, and biological degradation.

Chemical oxidation can be one of three types: atmospheric oxidation, photo-oxidation, or oxidation catalyzed by materials present in the oil. Alkene, aromatic, and alkane hydrocarbons with suitable side chains will be attacked most readily. However, the requirements in terms of dissolved oxygen for the oxidation of oil make the process fairly insignificant as compared to other weathering processes.

Evaporation affects low boiling compounds. It results in a selective loss of the low molecular weight compounds. The rate of evaporation is dependent primarily on the vapor pressure of the oil, but it is enhanced by high winds, rough seas, high sea temperature, irradiation, and increased surface area. Some components of crude oil evaporate much more slowly than others, forming an "atmospheric residue" on the surface. The residues have higher specific gravity and viscosities and greater concentrations of sulfur and inorganics than the fresh oil. The residue may even develop a specific gravity greater than sea water, and sink. This does not normally occur, however, unless the oil is occluded by material of a high specific gravity, such as silt or clay.

Dissolution preferentially affects lower molecular weight hydrocarbons, with aromatics dissolving more readily than alkanes of the same boiling point. Dissolution is complemented, to a degree, by the formation of emulsions. Oil-in-water emulsions have the appeal of removing oil from view; however, they do not seem to form without the aid of a surface active agent. Water-in-oil emulsions, on the other hand, do form naturally at sea, producing the characteristic "chocolate mousse." These emulsions are hard to disperse and slow to degrade.

Biodegradation has been extensively studied since the early 1900s. Bacteria originally considered a nuisance for attacking refined petroleum products seemed particularly useful when the oil became a nuisance. Recent research has been directed toward finding types of bacteria and other microbes demonstrating the capacity to degrade oil in both the presence and absence of oxygen, toward studying rates of degradation, and toward determining toxicities of degradation products on other organisms, particularly fish. In general, bacterial degradation affects homologous series of hydrocarbons at roughly the same rates. Normal alkanes are attacked preferentially, and this manifests as a lowering of ratios between straight chain and adjacent branched alkanes (19). Extended degradation seems to show a serial preference for branched alkanes, then cyclo-alkanes, and finally aromatics.

In combination with dissolution, biodegradation leads first to a rapid depletion of the soluble low boiling aromatics, followed in time by an increase in relative aromatic content of the oil residue as the less refractory saturated hydrocarbons are degraded. As a result, weathered oil may be more toxic than fresh oil of equal weight after considerable bacterial attack.

More than 100 species of bacteria, yeasts, and fungi have been shown to oxidize hydrocarbons (20,21). No single microbial species can utilize all hydrocarbons, rather each is limited in the scope of its nutritional capabilities. Generally bacteria seem to require normal physiological mineral salt solutions, nitrogen and phosphate sources, free oxygen, and almost neutral pH conditions. Some hydrocarbon oxidizers can utilize nitrate or sulfate rather than oxygen as hydrogen acceptors, but they are a small minority. Thus, biodegradation is generally much slower in the absence of oxygen.

Recently research has centered on the potential use of microbial seeding to clean up oil spills. Limiting factors in nature for the necessary rapid bacterial growth are nutrients (mainly nitrogen) and oxygen availability. Although degradation proceeds rapidly under laboratory conditions of optimal temperature and oxygen conditions, it seems that in nature bacterial degradation is a long-term process. Once oil is incorporated in anaerobic sediments (such as a few inches below the surface of most sediments), degradation appears to be extremely slow.

Chapter Three

Effects of Oil on Marine Ecosystems

Even though the old adage that oil and water do not mix is partially untrue, oil does tend to concentrate at the water's surface or, especially if absorbed in sediment, on the bottom. This means that the impact of oil on a marine ecosystem is not uniform but is greater on organisms living at or near the ocean's surface—intertidal life, neuston (small ocean-surface dwellers), and sea birds—and those organisms living on the seabed, the benthos. The following discussion is a summary of the knowledge of the effects of oil on some major marine communities and environments.

BIRDS AND MAMMALS

The layman's picture of the biological effects of oil spills is mainly of oil-fouled birds on beaches. The obviousness of this effect together with our feelings toward warm-blooded animals have meant that the oiling of marine mammals and especially birds has received much publicity and has been the subject of much literature (5,22).

Many dead sea birds have been observed after most spills of crude and heavy fuel oils. The death toll has usually been estimated by counting the number of oiled birds washed onto beaches, but these estimates often do not include dead birds not reaching shore. It has been estimated that only 5 to 15 percent of those birds killed by oil actually wash onto shores (4). Nonetheless, 40 thousand to 100 thousand birds were reported killed by the *Torrey Canyon* spill (12), 3,686 by the Santa Barbara blowout (23), and 7 thousand by the San Francisco Bay spill (24). Such spills have often killed significant proportions of some bird species within the affected areas.

The primary effect of oil on birds has been the fouling of feathers, to which oil clings. The small feather barbules are disheveled, thus allowing a disruption of the insulation and buoyancy of the feathers. Birds may sink and drown or may rapidly lose body heat and succumb to pneumonia. Fouled birds exhibit excessive preening, which may disrupt normal feeding activity or result in ingestion of oil. Ingested oil may poison the birds through inflammation of the digestive tract or disturb other physiological processes. Birds in which the insulating ability of the feathers has been damaged develop a very high metabolic rate to compensate for the rapid loss of body heat. This, coupled with a cessation of feeding, may lead to "accelerated starvation." Oiled birds may not lay eggs as usual and oiled eggs may not hatch.

Some species of sea birds are more susceptible to oiling than others. Diving sea birds suffer inordinate mortalities, whereas others, such as sea gulls and shearwaters, evidently avoid oil. There have been reports that the more susceptible species are those that are attracted by the slick and land on it. But another important factor is that these diving birds, which may remain submerged for several minutes, may surface under oil slicks. In Western Europe, auks, puffins, razorbills, guillemots (called murres in North America) and some sea ducks have been especially hard hit by oil spills (5).

During the Santa Barbara incident, loons and grebes, which made up 7 to 10 percent of the total bird population, suffered 64 percent of the mortality. This means that a very large proportion (one half or more) of the loons and grebes in the Channel were killed (25). Many of these sea birds have long lives and breed slowly; thus recovery of the populations may be a very slow process. Because of this there has been concern expressed over the survival of European populations of certain bird species, mainly several species of auks. To complicate matters, because of migratory behavior, a large proportion of a bird species may often be concentrated in a small area; oil pollution could thus gravely reduce populations. In fact it appears that the jackass penguin currently faces extinction because the species is restricted to South Africa and is gradually being killed off by floating oil emanating from tankers rounding the Cape of Good Hope (26).

The treating of oiled birds has received much publicity, but heretofore has met with only limited success. Generally 20 percent or fewer of the treated birds have survived, and the survival record of those species most susceptible to oiling has been dismal. Many birds die soon after capture, others do not survive the cleaning procedures or the captivity during recuperation. Early attempts at treatment failed because of the use of harmful detergents and improper feeding and handling of "rehabilitating" birds. British units for the rehabilitation of oiled sea birds are now able to achieve nearly complete survival of treated birds (27), but this level of success is not possible without established facilities and trained personnel. The problems of treatment, together with the fact that only a fraction of oiled birds may reach shore and can be captured, cast doubt on whether rehabilitation of oiled sea birds could appreciably decrease the impact of a major spill on sea bird populations, except perhaps in Britain.

In summary, great destruction of sea birds is an undeniable consequence of oil pollution at sea. Sea birds are frequent victims—the only ones known with certainty—of the routine intentional discharge of oil by tankers at sea. Populations of several species face local extinction in the northeastern Atlantic, and one penguin species is imperiled with global extinction.

The effects of oil on marine mammals are much less well known than for birds, nonetheless they do not seem nearly as catastrophic. Marine mammals are not particularly abundant along most coasts of the world, but concern was expressed for the sea lions and elephant seals of Santa Barbara's Channel Islands, and for the migrating endangered grey whales. Despite confusing public reports, there is no conclusive evidence that any of these mammals were killed by the Santa Barbara blowout (28,29). This was in part fortuitous, since the elephant seal pups had been recently weaned and were not feeding. Had the spill occurred earlier or later, there is a greater probability that they might have ingested oil and so increased the chance of injury.

Mammals have been affected by some spills (4), however. Possible effects include suffocation by oil, disruption of insulation abilities of fur coats, and poisoning from ingested oil. In addition to pinnipeds (seals, walruses, etc.) and cetaceans (whales and dolphins), other mammals occupy coastal marine habitats. Sea otters are recovering from near extinction along the West Coast of the U.S. Commercially valuable fur bearers, mainly muskrats, inhabit coastal wetlands in Louisiana and are susceptible to oil pollution from the production fields, although fur production has not declined there (30).

FISHES

There are few reports of directly lethal effects of spilled oil on fishes. Tendron (31) has reported a decrease in the numbers of fish off Ushant in the English Channel after the *Torrey Canyon* spill. Also after that spill 50 to 90 percent of the eggs of pilchard were dead, and young pilchard were rare in plankton samples taken off the coast of Cornwall (12); but again the effects there were complicated by the use of dispersants. Fish kills were observed after a crude oil spill in Puerto Rico (32), a spill of a variety of fuel oils at Wake Island (33), and the fuel oil spill at West Falmouth (14). On the other hand a thorough study by Ebeling et al. (34) could detect no effect of the "Platform A" blowout on Santa Barbara Channel fish communities.

There is a large body of evidence that oil, refined products, and components may be quite toxic to fishes in laboratory experiments (4). Fishes may be more resistant than some other marine organisms because their surfaces, including gills, are coated with a slimy mucus that is oil repellent. Also, evidence suggests that fishes may simply move away from an area affected by oil (10,34).

There has been concern that larval fishes, which often concentrate at the ocean surface, may be adversely affected by floating oil, either through toxicity or entrapment (4,35). Numerous deaths among these young forms could have serious consequences for adult populations over the long run, and such consequences may be difficult to detect, although a recently developed physical model (36) suggests that they would be minimal. Also unknown are the long-term effects of trace contamination of fishes through feeding. Many fishes feed on benthic invertebrates that may contain petroleum hydrocarbons from the sediment. Others, such as mullet, feed on organisms and detritus at the water's surface, where there may also be hydrocarbon concentrations. Mullet in some Australian coastal waters close to port and refinery facilities are frequently contaminated with petroleum hydrocarbons having an odor and chemical composition similar to kerosene (37,38).

FISHERIES

In addition to killing some organisms outright, oil pollution may otherwise affect the use of commercial species of fish, molluscs, and crustaceans. Even though no direct effects of oil on fishes were observed in Santa Barbara Channel (34,39), the risk of fouling fishing gear or catching contaminated fish severely reduced fishing effort. This had an economic impact on the local fisherman, even though the resource was not directly affected (40).

Tainting of commercial species of fish (4,37,38), clams (41), oysters (9,42,30), and mussels (4,6) by oil has been frequently reported. This results in unsaleable catches. Tainting may be quite persistent, the noticeable taint lasting several months. Furthermore, Blumer et al. (19) have indicated that oysters affected by the West Falmouth spill, but subsequently kept in clean water, retained petroleum hydrocarbons similar in chemical composition to the Number 2 fuel oil spilled for over eight months. Based on these findings, authorities imposed a total ban on the taking of shellfish in a large area. A partial ban still exists more than three years after the original contamination. Ehrhardt (43), considering Blumer's findings, his own findings on heavy petroleum contamination of Galveston Bay oysters, and the kerosene-like tainting of Australian mullet (37,38), concluded that organisms accumulate, quite unspecifically, the entire range of hydrocarbons to which they are exposed.

More recent research indicates that shellfish (44,45) and fish (46) will accumulate hydrocarbons from solution but will rapidly (certainly within a few months) shed them if maintained in uncontaminated seawater. Blumer's conclusion that petroleum hydrocarbons will persist without much modification or reduction for many months in the fatty tissue of shellfish thus seems oversimplified. The mechanisms of uptake and of decontamination are complex and varied and the time required for decontamination probably depends on the duration and dosage of exposure, on storage sites, on whether the species involved possesses the metabolic processes to detoxify the hydrocarbons, and on the organisms' physiological state and other factors.

Attempts have been made to demonstrate the effects, or lack

thereof, of the release of petroleum on fishery productivity as reflected in commercial catch statistics. Notable among these was a study undertaken along the coast of Louisiana, which supports very large and important commercial fisheries as well as a highly developed petroleum industry. Statistics show no decline in catches of shrimp, crabs, and fish concommitant with the development of the petroleum industry (30). A decline in the oyster harvest coincident with the expansion of the petroleum industry in the 1940s has been attributed to an oyster disease and not oil. While it is tempting to seize upon this as evidence that oil is not detrimental to the productivity of a large coastal area, several shortcomings in fishery catch statistics need to be pointed out. They do not represent precise reporting, because they fail to take into account the changing effort and technological advances of fisheries and are generally not available for the localized areas in which pollution may be intense. On the other hand, field surveys in Texas indicate that the production of fish and shrimp is reduced in waters polluted by large oil-producing operations (47).

It is certainly apparent that the long history of oil production on the Gulf Coast has not resulted in large scale devastation of fisheries, but data are inconclusive on the possibility of important local effects.

PLANKTON

Few observable effects of oil spills on the small, passively drifting plants and animals composing the plankton have been uncovered in post-spill investigations. Some kills were observed during the *Torrey Canyon* spill among phytoplankton (plant plankton), but none among the zooplankton (animal plankton) (12). In any case, the heavy use of chemical dispersants complicates the issue. Studies following the Santa Barbara blowout could detect no harmful effects on phytoplanktonic productivity (48) or zooplankton populations (49). However, because plankton is passively carried about by water currents, it would be very difficult to discern effects in the field, especially in open waters like the English or Santa Barbara Channels.

Laboratory experiments have generally produced more tangible results, although the degree to which these results may be extrapolated to natural conditions is open to question. As early as 1935, Galtsoff et al (50) found that growth in cultures of diatoms, which are important components of the phytoplankton, was inhibited by high concentrations of oil. Russian workers have also found that various diatom species are sensitive to kerosene and fuel oils (51). Recently, researchers have found that water extracts of various crude and fuel oils and dispersants may inhibit growth of phytoplankters (52,53,54,55), or affect the rate of photosynthesis (53,54,56). However, it is difficult to relate the results of most of these studies to what might actually be experienced in the marine environment, because either the actual concentrations of hydrocarbons in the experimental water was unreported or the concentrations used in the experiments are unrealistically high. Gordon and Prouse (56) did use experimental concentrations that realistically might be found in polluted coastal areas or near spills. They found that, at concentrations of petroleum hydrocarbons in solution below 10 to 30 parts per billion (ppb), photosynthesis was stimulated. Concentrations of the crude and refined oils tested of from 60 to 200 ppb suppressed photosynthesis to varying degrees. Number 2 fuel oil had the greatest effect and depressed photosynthesis by 60 percent at concentrations between 100 and 200 ppb. These results indicate that there is a possibility of both stimulation and inhibition of photosynthesis in areas subject to chronic oil pollution or in the vicinity of a heavy oil spill.

The larvae or young of many benthic and fish species spend time as members of the zooplankton. They are often much more susceptible to toxicants than adults. Larvae of the intertidal barnacle (*Elminius modestus*) were shown to be killed by 100 parts per million (ppm) of fresh crude oil (57). Crude oils have also been shown to be toxic to the planktonic eggs and larvae of some fishes, including cod and herring (58).

In addition to potentially acute effects of oil spills on planktonic organisms, there has been concern about the long-term effects of floating oil and tar lumps, which have become alarmingly common on the high seas. If the concentrations of petroleum hydrocarbons in ocean surface waters are being increased by shipping discharges or atmospheric input, there would certainly be concern for the near-surface plankton so important to oceanic productivity. Data on hydrocarbon concentrations in seawater are remarkably scant and historical data are nonexistent, thus making it impossible to predict future trends and effects of oil on the oceanic ecosystem.

NEUSTON

A unique, but poorly studied, assemblage of organisms lives right at or just below the surface of the sea, the neuston. Because of their intimate contact with floating oil, it is difficult to imagine that neustonic organisms would not suffer toxic or mechanical effects of contact with fresh oil slicks. Concern has been expressed for the community of highly adapted organisms associated with *Sargassum*, which floats over much of the North Atlantic. It is significant to note that petroleum hydrocarbon contamination of *Sargassum* plants and animals has been reported (59). Some research is now in progress in Bermuda on the effects of floating oil on the *Sargassum* community. Unfortunately, the ecology of neustonic organisms is very poorly known, and the effects on it of floating oil can only be surmised.

INTERTIDAL ORGANISMS

Spilled oil has its most visual effects on the intertidal environment. The obvious effects plus the special attention that has always been given intertidal organisms by marine biologists have meant that these creatures have often been the only ones studied after a spill.

Oil may smother, foul, or directly poison, intertidal organisms. Reports of the effects of direct oil poisoning vary from "non-existent" to "extremely damaging." The differences seem in part due to the type of oil spilled, which determines both the toxicity and the degree of of contact with fresh oil slicks. Even crude oils differ in their toxicities to intertidal organisms (34). The physiology of toxicity and the hydrocarbon components directly responsible for it have not been adequately studied.

The physical effects of oil–especially notable with crude oils, highly weathered oils, and heavy fuel oils (Bunker C)–include smothering, abrasion, removal, and alteration of the substrate. Organisms show varying resistance to smothering. At Santa Barbara one sessile barnacle species suffered high mortality because individuals were completely covered, while another taller species projected above the oil coating and fared much better (61,62).

Organisms having a natural coating of mucilage or mucus, such as many macroalgae and anemones, are often unaffected by oil that does not adhere to their surfaces. Although the slick algal species, including the important giant kelp, fared well after the Santa Barbara spill, intertidal surf grass, a vascular plant without a mucilage coating, suffered great losses (15).

Large algae are important intertidal primary producers on rocky shores. Although generally resistant to oil, algae may sometimes be coated with oil and torn from the substrate. Oil taken up or stuck to algae may be passed to grazing invertebrates (63). Even though algae may be initially affected by spilled oil, their mode of reproduction and rapid growth often allow them to rapidly recolonize a polluted shore. Particularly if herbivorous invertebrates are killed, green algae (e.g. *Enteromorpha* and *Ulva*) will rapidly recolonize denuded intertidal rocks, producing a characteristic "green phase" (4). This will gradually be replaced with a cover of brown fucoid algae, which are more robust. If the fucoids are not grazed they may form a thick cover that inhibits the reestablishment of the usual fauna of mussels and barnacles. The recovery of herbivore populations and consequent reestablishment of an algae/grazer balance are necessary for the complete recovery of the system. In some instances the heavy algal cover has provided a relatively protected environment, allowing the development of more diverse and abundant fauna than before the pollution.

Intertidal animals may be killed in a number of ways. Most investigations have shown that direct toxicity is rather rare. The most devastating toxic kills have occurred after the *Tampico Maru* spill on a rocky shore (10) and at West Falmouth (38) on a sand and mud shore. Both of these spills were of light fuel oils and were concentrated in small areas. At the site of the *Tampico* spill almost all intertidal organisms were killed; surviving species were high intertidal, spray zone dwellers or particularly resistant anemones. At West Falmouth mortality was also nearly complete, and only a few resistant polychaete worms remained.

More usually reported are non-toxic lethal effects on intertidal organisms; this is especially the case in crude or heavy fuel oil spills. Some mobile species such as snails retract into their shells and thus lose their purchase on a rocky shore when exposed to oil. These animals may then be swept away from their intertidal habitats by waves. Still other molluscs may become so encrusted with oil that they can no longer hold on to their substrate and hence are washed away. Some mobile intertidal animals may be cemented to a rock's surface by particularly thick oils (65). Sessile animals, such as barnacles, may be covered and suffocated by oil deposits. This was the cause of large barnacle mortalities on shores affected by the Santa Barbara (61,62) and San Francisco (24) spills.

Most post-spill investigations have concentrated on exposed rocky intertidal habitats, and hence the effects of oil on sand beach fauna remain something of a mystery. This is probably due to the difficulties in sampling and the tremendous natural variability in populations of sand beach animals. This has often led the investigator to abandon such studies or to conclude that no effects were discernible. Sheltered intertidal muds and sands have also been neglected by researchers; however, there is reason to believe that floating oil may have a great impact on such environments. Number 2 fuel oil spilled at West Falmouth (64) and Bunker C spilled at Chedabucto Bay (65) had severe effects on sheltered intertidal animals. Oil at both sandy locations persisted in the intertidal sediments for several years, whereas on exposed rocky shores the intertidal zone is usually cleansed of oil within a short period by high wave energy.

On rocky shores stranded oil may weather to an asphaltic pavement on the surface of upper intertidal rocks. These deposits erode slowly and may persist for several years. Barnacles may set on these asphaltic surfaces, but there is some evidence that their chance of survival may be low, because their black substrates absorb more heat than the lighter rocks, thus increasing heat stress and dessication. Intertidal asphaltic surfaces have been recolonized by barnacles on Santa Barbara shores (66), although such recolonization was not apparent in the San Francisco Bay area (24) until the deposits were worn away.

Chronic exposure to oil has been shown to be more harmful than one acute dosage. Crapp (67) has described the rocky shore community adjacent to a refinery outfall in Milford Haven, Wales. Along a limited stretch of shoreline adjacent to the outfall, the community was dominated by fucoid algae and limpets were rare or absent, whereas on adjacent shores barnacles and limpets were dominant and the growth of fucoids much reduced. The effects of chronic exposure to crude oil from natural oil seeps off Southern California have been a matter of controversy. Initial reports (61) indicated that the diversity of biota was lower on shores exposed to seeps, but reevaluation (66,68) indicated that there is no sound evidence for this contention.

In summary, oil spills have had considerably varied, but generally slight to moderate effects on intertidal organisms, and thus perturbed communities have recovered within a few years. This resistance and resilience is due to (1) the hardiness of intertidal organisms, (2) their rapid reproduction, and (3) the relatively rapid removal of oil from the intertidal zone by waves. The use of cosmetic clean-up techniques–such as toxic dispersants and steam cleaning–on oiled shores often does more harm than good and should be stringently controlled and generally discouraged. Most biologists who have experience with the effects of such techniques recommend no attempt at removal unless clearly necessary, and then only with absorbents or by scraping away oily sand.

SEABED ORGANISMS

Oil has a distinct affinity for sediment particles, especially clay minerals and particles coated with organic matter. Thus, much of the oil spilled at Santa Barbara met with water masses made turbid by heavy land run-off, formed denser-than-water, oil-sediment aggregates, and sank to the bottom of Santa Barbara Channel (69). Subtidal bottoms were the ultimate depository of much of the oil spilled at Santa Barbara (70), at West Falmouth (14), in the English Channel (12), and in Chedabucto Bay, Canada (71).

While the British worsened the littoral pollution from *Torrey Canyon* through excessive application of toxic detergents, the French sank the oil before most of it reached shore, thus eliminating obvious intertidal pollution, but doing potentially serious if not obvious harm to bottom life. Unfortunately, the seabed organisms' subtidal habitat, especially soft, sediment-covered bottoms, has been largely ignored in most post-spill studies. Instead, concern and research effort has concentrated on the intertidal habitat.

Observations on the subtidal effects of oil spills were made on larger benthic (seabed) organisms by skin divers after the *Tampico Maru* and *Torrey Canyon* spills. Extensive kills of Pismo clams, abalones, starfish, sea urchins, lobsters, and other subtidal benthic invertebrates occurred within the small cove into which *Tampico* oil was spilled (10). The persistence of oil in the cove and the ecological imbalance resulting from the reduction of grazing invertebrates (mainly sea urchins) and predators altered the ecology of the cove such that complete recovery could not be reported seven years after the spill.

Large numbers of dead and moribund clams, snails, crustaceans, and echinoderms were noted by divers just off shores oiled by the *Torrey Canyon* spill (12). The effect of oil itself on the benthic organisms is impossible to assess because of the overwhelming effects of toxic dispersants applied to nearby shores. Approximately 23,500 tons of *Torrey Canyon* oil was sunk off the coast of Brittany by the application of 3,000 tons of chalk in order to prevent it from reaching shore (12). Virtually nothing is known about the effects of this oil on bottom life or its persistence in the bottom sediments. Large masses of oil on the bottom have subsequently fouled fishing gear and contaminated catches off the French coast (4).

Detailed quantitative sampling of less obvious benthic animals living within the sediment was undertaken after the Santa Barbara blowout (72) and after the West Falmouth oil spill (64). A large scale survey of benthic animals had been undertaken ten years before the Santa Barbara spill, so there was some background upon which to judge the effects of the oil spill. The weight of living organisms (biomass) found after the spill was considerably lower than that ten years previous. This was especially evident in the dense beds of the echiuroid worm *Listriolobus pelodes*; such beds are not found outside of the Santa Barbara Channel area. Fauchald (72) has surmised that the observed reductions in population levels could have been due to heavy rainfall around the time of the spill, the effects of drilling and spills in the channel, the increasing pressure of land-derived pollutants (for example, pesticides, and sewage), or even natural fluctuations in population levels.

Unfortunately the quantitative animal samples taken remain incompletely analyzed. The small crustaceans so susceptible to oil pollution at West Falmouth (64) were not studied. It is impossible to determine from the data presented, given the limitations of sampling design and the time elapsed since the last survey, whether subtle changes in the structure of the benthic communities of the Santa Barbara Channel had occurred, much less whether any changes were due to the oil spill. It is significant to note here, however, that Kolpack et al. (70) reported that large quantities of oil were deposited in the bottom sediments of Santa Barbara Channel. These oil-laden sediments were later transported by bottom currents and ultimately deposited in the Santa Barbara Basin, where degradation is extremely slow because of low oxygen conditions (73). The net effect has been that oil has spread along the bottom from the originally contaminated area near Platform A to contaminate most of the bottom of the Channel.

The catastrophic effects of Number 2 fuel oil on subtidal benthos in the West Falmouth area have been documented by H.L. Sanders and his colleagues (14,64,74). Initially, massive kills were indicated when large numbers of dead bottom animals, including lobsters, were washed ashore. Evidence indicated that the oil pollution spread from shallow areas to deeper areas over the course of the year following the spill, thus lethally contaminating new areas. As in Santa Barbara, the mode of transport seemed to be the movement of oil-laden sediment along the bottom.

In the most heavily oiled bottoms, the animals were nearly totally eliminated; in the peripheral areas, the more susceptible forms such as amphipod crustaceans were selectively killed. Recolonization of polluted areas was mainly by hardy opportunists, such as capitellid polychaete worms–marine "weeds"–which developed large populations, taking advantage of reduced competition and predation. Now, nearly four years after the spill, significant areas of bottom remain grossly polluted, although the fauna is recovering throughout most of the area (75). Estimates by Woods Hole scientists of the amount of time needed for reasonably complete recovery range from five to ten years after the spill.

Scarratt and Zitko (76) found that even though no lethal effects were apparent in subtidal benthos in an area contaminated by the spill of Bunker C fuel at Chedabucto Bay, many animals were still contaminated twenty-six months after the spill, having petroleum hydrocarbons with chemical "finger prints" like that of the Bunker fuel.

The effects of oil pollution on subtidal seabed organisms have been seriously neglected. Yet oil is often deposited in high concentrations on the bottom, where it may persist and chronically repollute an environment. Biodegradation and other weathering processes result in a selective loss of n-alkanes so that the relative composition of the oil persisting in sediments changes markedly as the total quantity of oil is reduced (19). The net effect is that the hydrocarbons remaining are rich in aromatics and cycloalkanes which may continue to be harmful.

Marine organisms may accumulate petroleum hydrocarbons in their tissues (42,43,44,45). This may be especially true for benthic organisms, many of which feed on suspended matter or bottom sediments which may contain oil. The chronic effects of such contamination remain unstudied, and it is unknown if petroleum hydrocarbons can be transmitted to fishes feeding on oil-contaminated seabed organisms.

WETLANDS

Tidal wetlands are characteristic of many estuarine shores throughout the world and include salt marshes dominated by grasses in temperate climates, and mangrove swamps dominated by trees in the tropics.

The great value of wetlands in coastal ecosystems is generally accepted, if not always well understood. They serve as habitat, feeding, or nesting grounds for shore birds, fish, and other wildlife. Tidal wetlands have been shown to be among the most productive environments on earth, and it is this great productivity that supports much of the life in estuaries through a food web based on vascular plant debris (detritus). Wetlands also play a considerable role as geological agents and are important to shoreline stabilization.

Salt marshes have generally proven to be resistant in the face of many types of environmental onslaughts. In the case of oil pollution, they have often suffered only minimal damage and have also alleviated the pollution problem by trapping and holding oil. The frequent proximity of marshes to sources of chronic hydrocarbon pollution has given impetus to both post-accident and experimental studies of the effects of oil. The first concerted studies took place in Louisiana marshes and were sponsored by the oil industry in the face of charges by fishermen of pollution from drilling (77,78). These studies involved experimental application of oil on marsh plants with an assessment of changes in their biomass after treatment. The experiments indicated that a moderate dosage of oil was not excessively harmful, but that repeated applications proved lethal. Oiling also apparently produced a "fertilizing" effect of stimulated growth.

Salt marshes polluted by the *Torrey Canyon* (79), *Chryssi P. Goulandris* (80), West Falmouth (63), and *Arrow* spills (65) have also been investigated. Baker's extensive field experimental work (81) has also shed considerable light on the effects of oil on marsh plants. Most of these studies have shown that marsh plants survive light to moderate oiling in a single dosage. Immediate effects include death of heavily oiled shoots, followed by regrowth from living roots. Multiple dosing, however, can do considerable damage to marsh plants.

The amount of damage seems to be determined by the type and amount of oil, the plant species involved, and the time of year. Weathered oil with few aromatics is less toxic than fresh oil or light fuel oils. Very heavy oiling can smother marsh plants. The weathered oil reaching and affecting marsh plants in Brittany after the *Torrey Canyon* spill was an impermeable barrier to gas exchange rather than a direct poison (79). This lack of gas exchange may also lower the oxygen levels of the marsh soil (82).

The effects of direct contact of oil on plants has been extensively studied because of the use of petroleum products in herbicides and insecticides (83). Hydrocarbons can penetrate into plants through stomata (pores which permit gaseous exchange), or in the case of toxic oils, directly from the point of contact. Once in the plant, oil can travel within intercellular spaces. Oil reduces transpiration and translocation, probably by blocking these spaces and/or stomata. Respiration rates of plants may be altered (increased or decreased) by oil's effective disruption of metabolic mechanisms. Plants may also have mechanisms, either epidermal or cellular, to resist penetration by oil.

The effect of oil on marsh plant species depends on seasonal growth processes. Oiling during the growing season may cause considerable damage, but not at other times. Oil may also influence flowering, seed development, and vegetative reproduction by underground roots. If oiled during a growing season, annual plants may suffer more than perennials, which may survive by regeneration from roots.

On several occasions, investigators have noted apparent growth stimulation following oiling (77,84). Exact causes for stimulation are unknown, but possible explanations include: (1) increased water retention of oiled soil, (2) release of nutrients from oil-killed animals, (3) plant nutrients or growth regulators in oil itself, and (4) nitrogen-fixation by oil-degrading microorganisms.

Most of the observations on the effects of oil on marshes have been confined to the marsh plants themselves. The West Falmouth spill killed not only saltmarsh cordgrass and other marsh plants, but also saltmarsh animals, including molluscs and fiddler crabs (63). Burns and Teal (63) analyzed marsh plants, animals, and sediments for petroleum contamination more than a year after the initial spill. They found that all the organisms examined at the site were contaminated, but that organisms from nearby non-polluted areas were not. These organisms included two algal species, an annual plant (*Salicornia*), a perennial (*Spartina alterniflora*), a mussel, the killifish *Fundulus*, the American eel, and the herring gull. Oil was detectable as deep as 70 cm in marsh sediments, and this is in marked contrast to previous observations that marsh sediments were uncontaminated by oil or that oil was restricted to the upper 5 to 10 cm. Burns and Teal further noted that there had been a reduction in straight chain alkanes in relation to branched chain and cyclic alkanes and aromatics since the original pollution, and that although oil was obviously taken in by organisms, there was no evidence of food chain magnification.

Repetitive or chronic oil pollution as found near an oil terminal or refinery effluent can have a decidedly harmful effect on salt marshes. Baker's experiments (81) indicate that more than two or three fresh oil spills a year or one or two light oilings per year with weathered oil for many years would severely reduce marsh grasses. Much of a marsh near a refinery effluent in Southampton was completely denuded by the chronic low-level pollution from the effluent (85). Apparently the grasses were killed by being constantly coated with thin oil films. The loss of grass cover resulted in rapid erosion of the marsh bank.

Chronic pollution of salt marshes can have obviously serious effects, but the effects of a single oiling may be quite variable, depending in part on the factors discussed above. Although most studies have not uncovered significant effects, oils as different as the light, aromatic-rich Number 2 oil spilled at West Falmouth, and the heavy, less toxic Bunker C lost from the *Arrow* are clearly harmful to marsh plants and animals. The effects of oil on marsh organisms other than vascular plants have generally not been studied. Clearly, further investigations should consider the smaller, less obvious, but nonetheless integral organisms of the sediment surface. Also, the findings of Burns and Teal make chemical analysis of organisms and sediments advisable.

Mangrove swamps are not as well studied as salt marshes. Consequently, we lack the necessary basic and practical information to make conclusions about the effects of oil on these environments (see 5). Spills in Puerto Rico (32) and in Panama (86) have killed mangrove trees and associated animals. Mangrove swamps will conceivably suffer more pollution as offshore oil production in the tropics increases.

SOME SPECIAL ENVIRONMENTS

It is obviously unrealistic to suppose that an oil spill will have the same ecological consequences anywhere in the world. Although a systematic geographical analysis is clearly beyond the scope of this report, three "special" marine environments deserve attention because of their importance, and because of the increasing threat of oil pollution in these environments: polar regions, tropical coral reefs, and estuaries.

I. Polar Regions

The tapping of oil resources in the North American arctic and the environmental consequences of the Trans-Alaska pipeline have received much public attention. Ultimately Alaskan oil will be transported at sea, from a southern terminus if not from northern sources directly. Because of the hazards to navigation in these waters, there is a significant probability of oil spills in these polar marine environments. There are reasons to believe that the effects of spilled oil in polar regions might be serious and long lasting (87), including: (1) cold temperatures do not permit rapid evaporation of aromatics in oil, thus allowing more of these toxic hydrocarbons to enter solution in sea water even though the solubility of these compounds is lower at low temperatures; (2) the rate of bacterial degradation and other processes of weathering are comparatively slower at very cold temperatures; and (3) the marine biota of polar regions are generally long-lived, have low reproductive potentials and do not have wide ranging dispersal stages (88).

These combined indicate that a large oil spill in the North American arctic might have considerable immediate effects, that the oil would persist and remain toxic for a long time, and that recovery of an affected area through recolonization would be very slow. These potential impacts should be considered in light of the relative inefficiency of spill prevention and control techniques in these difficult environments. These observations are admittedly conjectural, and hopefully more light will be thrown on the biological impact of polar oil spills through ongoing investigations by Alaskan and Canadian ecologists.

II. Coral Reefs

Recently the threat of oil pollution to coral reef environments has become real. Oil tankers must travel extensively through tropical waters, especially off the coasts of Africa and Asia. Offshore oil drilling has commenced in the vicinity of Indonesia, and exploration has begun on the Great Barrier Reef. What would be the ecological consequences of spilled oil on the coral reef ecosystem, an environment of incredible diversity and productivity? The Australian government appointed a Royal Commission to investigate the consequences of an offshore oil industry in Great Barrier Reef waters (89). They heard extensive testimony on the ecology of coral reefs and the efficiency of cleanup operations, but there was a paltry amount of information available on the effects of oil on coral reefs (5).

Most observations made either during accidental spills or during experimental exposure of corals to oil offer no conclusive evidence that floating oil damages corals. However, Johannes et al. (90) found in field experiments that some corals, especially branching varieties, are seriously damaged if coated by oil while exposed to air. This finding is signifcant, since actively growing portions of many reefs, especially those in the Pacific and Indian Oceans, protrude well above the water's surface at low tide. Corals secrete copious amounts of mucus, which may serve to protect the individual coral polyps from direct contact with oil. On the other hand, an important and living part of a coral reef community is based on dead coral rubble, which is porous limestone capable of absorbing and holding oil. Although weathering and degradation of oil in tropical environments may be relatively rapid, the best scientific evidence indicates that recovery of a decimated coral reef would be an extremely slow process.

III. Estuaries

Estuaries are extremely productive and valuable ecosystems. They are often subjected to the intense pressure of multiple uses by man–expected to assimilate domestic and industrial wastes and accommodate marine transportation, yet continue to remain productive in fishes and shellfishes. Because of the increasing trend toward tanker-transported crude and refined oils and the proximity of estuaries to population centers, estuaries are convenient sites for oil receiving ports and refineries. Many estuaries must cope with increasing levels of petroleum pollution from accidental spills, ships, and industrial and domestic effluents. Thus, certain estuarine environments may be subjected to frequent small spills, often of more toxic refined products, or continuous low-level additions from refineries, petrochemical plants, oil field wastes, and even domestic sewage and urban runoff.

Probably because the oil spills that have taken place in estuaries have generally been small, their biological effects have generally not been well studied. A few broad observations may nonetheless be made based on what we know about the behavior of oil in the environment and the characteristics of estuaries. Several characteristics of estuaries suggest that oil pollution may have serious effects there. Because estuaries are generally confined and relatively shallow bodies of water, oil spills may not spread over a large area of the water's surface and have little chance to be swept to sea. Instead, there is a high likelihood that the oil will reach shore or the bottom. Estuaries are typically turbid, and therefore floating oil may tend to absorb onto fine sediment particles and sink to the bottom, where it may kill or contaminate bottom-dwelling organisms, including shellfish and bottom-feeding fishes. Indeed, this process seemed to be the principal cause of the widespread contamination following the West Falmouth oil spill (14). The oleophilic nature of detritus, which is typically abundant in estuaries, increases the probability of ingestion of oil by estuarine detritivores. If oil is deposited in sediments, it may persist for long periods under the anaerobic conditions typical of subsurface estuarine sediments. Also, long-term and rather high-level contamination of sediments may result from continuous low-level inputs (industrial and domestic sources). The serious biological effects of sediment contamination in Los Angeles-Long Beach Harbor have been described by Reish (91).

On the other hand, we know that the microbial biota of estuaries is abundant, thus indicating that at least aerobic degradation is rapid. Also, assuming that the pollution is strictly acute (that is, no persistent contamination results), recovery of damaged communities should be rapid because of the great reproductive potentials typical of most estuarine organisms. Clearly this is one aspect of the oil pollution problem where full assessment of the biological problems depends on sound information on the fate of the introduced oil, particularly on the degree to which oil finds its way into estuarine sediments and on the changes in amount and composition of oil in sediments.

The intertidal areas of estuaries are often characterized by extensive tidal wetlands–salt marshes in temperate latitudes and mangrove swamps in the tropics. These are thought to be in large part responsible for the very high productivity of estuarine environments and are a mainstay of the detritus-based estuarine food chain. Wetlands are vulnerable to dosage by floating oil. Although most experimental evidence shows that marsh grasses suffer little from a single dosage of oil (92), oils as different as the light Number 2 fuel oil and the heavy Bunker C fuel oil have caused lethal damage to marsh plants at West Falmouth (63) and Chedabucto Bay, Canada (65). Furthermore, chronic pollution–such as in the vicinity of a refinery effluent or near an oil handling facility–can kill off marsh plants and bare marsh sediments to erosion (81,85).

There is a reasonable basis for concern, yet any statements about the effects of oil on estuarine productivity are, at this stage, strictly conjectural–which underlines the crying need for further research.

Chapter Four

Effects of Cleanup Techniques

Various measures are taken to lessen the visual and ecological impact of spilled oil. At least for most larger spills these measures have proved relatively ineffectual and in some cases have worsened the biological impact of an oil spill. Techniques have variously involved containment by barriers and physical removal of floating oil, the use of absorbent material to concentrate oil, sinking, burning, chemical dispersal, and steam cleaning of oiled shores (4,93).

Containment and removal are most desirable from the viewpoint of avoiding biological damage, because oil is removed from the environment without the addition of any foreign substance. Unfortunately, the floating booms and skimmers used are only efficient in calm water, and the technology developed to this point has not proved successful in severe weather. Another drawback of these and other techniques involving physical removal is that, to be effective, they must be applied immediately after the accident. The use of absorbents has been in favor in North America. Straw was used with some success following the Santa Barbara and San Francisco spills, both on floating and stranded oil. The ecological impact of the straw appears slight; however, dried oil and straw deposits in the intertidal appear to be more persistent than those of oil alone (24). Peat moss was used effectively to cleanse shores oiled by the *Arrow* in Chedabucto Bay. The use of synthetic absorbents (oleophilic plastic polymers) seems to be increasing, thus posing the problem of the persistence of the non-biodegradable absorbent material escaping recovery.

The sinking of oil with chalk or sand has often been recommended and used in Europe. However, from the biological viewpoint it appears to be among the least acceptable countermeasures. Sinking an oil slick may save the intertidal zone from pollution, but it deposits oil over a large area of the bottom, where it may persist in the sediments. In coastal and estuarine environments, it is the productive benthic life that supports most of the finfisheries as well as shellfisheries.

Burning appears biologically innocuous. However, technological difficulties are great and the application of wicking agents or oxidants are often required to sustain combustion. Burning only seems feasible on small, confined slicks.

Following the *Torrey Canyon* spill British authorities countered the fouling of beaches with a massive application of chemical dispersants. The dispersants are solutions of surfactants, which break oil down into droplets that enter water to form an emulsion. The biological damage resulting on treated shores was much greater than that due to oiling alone (12), and subsequent field and laboratory research (94,95) has shown that many dispersants are more toxic than the oil they are meant to remove. Furthermore, dispersant and oil mixtures are often more toxic than the dispersant alone. The great toxicity of these "first generation" dispersants comes from the solvents used as a carrier for the surfactant component. Newer dispersants, such as Corexit and BP 1100, are far less toxic than their predecessors (96); however, dispersant-oil mixtures may still be quite toxic.

More basically, dispersal of oil is often simply a cosmetic treatment and injects oil more firmly into the environment rather than removing it. Emulsification of oil may increase the amount of toxic compounds in solution and spread this pollution throughout the environment. There may be contingencies in which the use of chemical dispersants should be considered in order to prevent floating oil from damaging bird or sea mammal colonies. For example, contingency plans have been developed in the United Kingdom that provide for the dispersal of oil at sea slicks during certain times of the year and in restricted areas adjacent to important or endangered bird colonies (27). Only in cases of such reasonable environmental trade-offs, or in cases of fire hazard or restoration of important amenities should the use of dispersants be permitted, and then only under strict supervision and control.

Oiled shores in Britain and in California have been cleansed with steam. This has virtually exterminated intertidal life, thus doing more harm than good. The consensus of shore ecologists now seems to be that physical removal by the use of absorbents is the only relatively non-destructive technique for cleansing oiled rocky shores, and that dispersants or steam should not be used (4,66).

Oil has been removed from sand beaches by scraping off the oily sand and either disposing of it or removing the oil from the sand in separators (97). While this almost certainly does some short-term biological damage, the damage may be outweighed by the benefit of removing potentially persistent contamination.

Chapter Five

Long-Term Effects

Although there is little comprehensive documentation of long-term effects of acute or chronic oil pollution, possible long-term effects are diverse and complex, but may be roughly classified as (1) those related to the long-term recovery of a polluted ecosystem, (2) those due to chronic pollution (continuous or frequent inputs), and (3) those due to persistent contamination of ecosystems by petroleum hydrocarbons.

ECOSYSTEM RECOVERY

As there is disparate evidence on the acute effects of an oil spill, there are likewise varying estimates on the rates of recovery of disturbed ecosystems. At Santa Barbara, recolonization of affected rocky shores commenced within a month after dosage (98). The intertidal biota now appears normal except in certain areas of heavy asphaltic deposits (66). Likewise the rocky shore fauna affected by the oil spilled in the tanker collision at San Francisco seems to have recovered within a year or so (24,99). On the other hand, benthic, intertidal, and marsh biotas remain affected at West Falmouth, even though the time for recovery has been about the same. Although the latter situation seems to be attributable to the persistence of oil in this environment rather than to a slow pace of biological succession, these examples do illustrate the variable nature of long-term recovery.

If the organisms of an area were severely depopulated but the oil did not persist within the system, recovery by the strictly biotic phenomenon of succession may be a slow process in the case of biologically highly "structured" communities. In natural communities one species may act to exclude or hold in check another through competition, predation, or grazing pressure. If the dominant species is removed or reduced below a certain level, the other species may then gain a foothold through atypical survival of the young and resist further biotic pressure. A familiar example from the land is the pattern of development of forests in the southeastern U.S. hardwood forests are the climax community, the end product of a successional sequence of tree communities. If a fire should destroy a forest, the area is recolonized by grasses and then pines, both of which require intense sunlight and cannot grow under a forest canopy. The forest gradually becomes dominated by pines, then by a mixed stand of large pines and young hardwoods until the climax hardwood forest is finally reached. The whole process of recovery to the pre-fire condition through biological succession may take more than a century.

In marine communities succession is generally a much more rapid phenomenon, but recovery through succession may still take several years. On British rocky shores there exists a balanced community of barnacles, mussels, rockweed (large brown algae), and algal grazers, particularly limpets. The grazers keep algal growth firmly in check, restricting the amount of rock surface covered with algae by grazing on algal sporelings. On heavily oiled and vigorously cleaned shores of Cornwall all the limpets were killed and within two or three months they became clothed with green algae, which are otherwise not abundant on these shores (4). During the following year, a vigorous growth of the characteristic rockweeds replaced this "green phase." Young limpets began to reappear but were unable to graze the large algae, so that an unusually dense cover of seaweeds persisted three to four years later.

Similarly, the *Tampico Maru* spill killed many sea urchins that, through grazing on sporelings, had prevented the development of extensive kelp beds prior to the spill (10). Following the spill the diminished grazing pressure allowed kelp sporelings to survive and dense kelp beds to form throughout the affected cove. The distribution of kelp in the cove did not return to its pre-spill pattern for six years.

The length of time necessary for successional recovery of an oiled marine community ranges from weeks or months to perhaps a decade. This will depend on the structural complexity of the community and the degree of initial damage. Of course, persistent contamination by petroleum hydrocarbons, frequent spills, or chronic pollution may hinder natural succession, appreciably delaying recovery.

CHRONIC POLLUTION

Marine organisms may be chronically exposed to persistent oil contamination of their environment, or continuously or frequently exposed to petroleum hydrocarbons from refineries, petrochemical plants, oil ports, or other waste discharges. The persistence and spread of bottom deposits of spilled oil at West Falmouth (19,100), Santa Barbara (70,73), and Chedabucto Bay (65,76) indicates that oil from a single spill may delay ecological recovery by continuous toxicity or by encroaching on and repolluting a recovering community. Such persistent contamination or frequent dosing may continuously or repetitively disrupt community succession and consequently keep a community in an immature, low diversity state.

The apparent large increase in the amount of floating petroleum residues (tar balls) on the high seas, particularly the Atlantic Ocean, has raised the specter of chronic pollution of the open oceanic environment. The surface of the ocean may also receive large amounts of petroleum hydrocarbons from atmospheric fallout in addition to what typically comes from tanker ballast (1).

There is a major knowledge gap about the effects of this oil on oceanic phytoplankters, larval fishes and other temporary plankters, and the unique neustonic communities, including the specialized *Sargassum* community of the North Atlantic. The virtually complete lack of information on the biological effects of oil in the open ocean make any conclusions about seriousness of effects premature.

Production of petroleum along the coast and offshore accounts for a sizeable input of oil into the marine environment through the release of "bleedwater" and drilling muds bearing petroleum hydrocarbons. Transfer of oil at dockside almost inevitably results in small but frequent losses through equipment failure and human error. Even in the exemplary oil port at Milford Haven in England, which loses a nearly infinitesimal fraction of oil handled, oil pollution is a chronic problem.

Oil refineries release substantial quantities of oil during normal operations. Although the concentration of oil in effluents is low, ranging for the most part from 10 to 100 p.p.m., the total quantity of oil released may be several thousand gallons per day. Petrochemical plants may release an even more exotic array of hydrocarbons having greater toxicity than most refinery effluents.

Surprisingly, very little research has been conducted on the effects of chronic inputs of petroleum on coastal communities. Much of the information available has been reviewed by Copeland and Steed (101) and Baker (102).

Refinery effluents may have a considerable impact on the benthic life in confined bodies of water where dispersion of the effluent is not rapid (102). For example, animals inhabiting sediments in Los Angeles Harbor that received large quantities of oil industry wastes were eliminated or limited to a single tolerant polychaete (91). The greatest effects were apparently due to the depletion of oxygen on the bottom by oxygen-demanding wastes that concentrated in the sediments. Also, saltmarsh plants were killed by a refinery effluent released in sheltered tidal creeks at Southampton, England (85). On the other hand, effluents released in more exposed waters with rapid dispersion seem to have had considerably fewer biological effects (102). In open bays the area in which benthic life has been noticeably affected by refinery effluents or bleedwater discharges was confined to a radius of several hundred meters (102,103).

Studies on phytoplankton (104) and zooplankton (105) of Galveston Bay, Texas, indicate decreased species diversity in the area near the Houston Ship Channel, which is heavily burdened with petrochemical as well as other toxic wastes. The effects of lowered salinity and other toxicants compound the picture, however, and the field evidence that chronic oil pollution affects planktonic communities is not complete. However, the experiments of Gordon and Prouse (56) indicate that photosynthesis in chronically polluted coastal waters may be affected.

Swimming animals may vacate an unfavorable area and thus avoid harm. Hence, fish may be absent or less diverse around refinery outfalls or bleedwater discharges (106). This may effectively stop fishery productivity in certain areas, or at least reduce it (47).

Among the shallow water ecosystems of the Texas coast those receiving oily wastes are characterized by lowered species diversity, large diurnal fluctuations in dissolved oxygen concentration, and sometimes near-anaerobic reducing conditions at the bottom (101). Community metabolism—the combined amount and relationship of photosynthesis and respiration of the whole community—fluctuates wildly. Under some conditions, both photosynthesis and respiration are depressed by highly toxic materials; under others, metabolism is stimulated due to the decomposition of waste products and release of nutrients.

The effects of oil inputs from such land-based sources as domestic and industrial wastes and urban runoff have received even less attention. Farrington and Quinn (107) traced the cause of high concentrations of petroleum hydrocarbons in sediments and clams in Narragansett Bay, Rhode Island, to domestic sewage effluents. Clams from contaminated sediments there showed signs of physiological stress and abnormal growth (108). Based on the analysis of sewage discharges in Southern California, Storrs (109) has estimated the total U.S. input of petroleum into the marine environment via sewage as 200,000 metric tons per year, an amount roughly equal to the worldwide loss from tanker accidents.

ECOSYSTEM CONTAMINATION

Evidence is building that marine ecosystems are widely contaminated with hydrocarbons of a petroleum origin. Preliminary results from baseline studies being conducted as part of the International Decade of Ocean Exploration (IDOE) have shown petroleum hydrocarbon contamination of plankton off the Louisiana coast, offshore fishes, *Sargassum* community members, suspended matter (seston), and surface microlayer and subsurface water samples from the Atlantic Ocean (110). An IDOE report also cites other evidence of contamination of biota and sediments, but these are from areas that have received oil spills (West Falmouth and Chedabucto Bay) or are chronically polluted by sewage or petrochemical wastes (Galveston Bay, Narragansett Bay, and Brisbane, Australia).

This evidence as well as the increasing reports of floating oil and tar balls at sea lead to the conclusion that much of the ocean is at least to some degree contaminated with petroleum hydrocarbons, even in areas outside those affected by spills and oily waste disposal. Although the release of oil into the oceans via natural seeps is certainly not new to geological history, the rate of addition of oil to the world's oceans due to man's activities is at least greater than the natural release rate (111).

It is not difficult to conclude that virtually nothing is known about the effects of this widespread, low-level contamination. This persistence of oil in marine organisms and their environment should surely be considered in analyses of the long-term impact of petroleum contamination.

Tainted fish and shellfish may retain a noticeable oily taste or odor up to six months after exposure to oil even when kept in clean water (4,9). However, the presence of a taint is not a definitive test for contamination because the odorous or distasteful portion of the petroleum may be shed while other contaminants are retained. Unfortunately, the persistence of contamination has not often been documented with rigorous chemical analyses. Although Blumer's (42) contention of a long persistence of petroleum hydrocarbons in shellfish has been challenged, it is apparent that marine organisms may maintain certain levels of contamination if their medium continuously or frequently contains low levels of petroleum hydrocarbons.

It has been suggested that hydrocarbons may be further concentrated through "biological magnification," a process whereby compounds may accrue up a food chain (2). This phenomenon explains the very high levels of chlorinated hydrocarbon pesticides in birds of prey. Although natural hydrocarbons may apparently be concentrated in this manner, there is at present no convincing evidence of food chain magnification of petroleum hydrocarbons, and internal concentrations depend on equilibria with concentrations in seawater (66).

The persistence of petroleum hydrocarbons in water and in sediments remains largely unquantified. The more refractory and less soluble portions of the Number 2 fuel oil spilled at West Falmouth still remain in subtidal and marsh sediments over three years after the spill (19,112). Similarly oil from the 1969 blowout persists in Santa Barbara Channel sediments, and cores of sediments in the central basin have revealed buried oil from natural seeps 25 thousand years ago (73). Indeed the long-term persistence of hydrocarbons, especially under anaerobic conditions, is the basis of petroleum formation.

Persistence is obviously a relative quality, and even the so-called persistent pesticides are subject to degradation. It seems quite probable that some oil pollutants remain in the environment for years, decades, or longer, depending on conditions. Knowledge of the fate of petroleum hydrocarbons and quantification of their persistence are urgently required before a balance sheet of petroleum hydrocarbon flux can be constructed. Such an approach is necessary in order to fully comprehend the scale of the problem of contaminated oceans.

Persistent contamination may have grave, if yet undocumented consequences. Suggested low level effects include disruption of photosynthesis, respiration, and other metabolic functions, disturbance of normal behavior patterns associated with survival, feeding, schooling, reproduction and development of young, and possible carcinogenic and mutagenic effects (2).

The literature on sublethal effects is scant. Experiments were often poorly conducted and experimental concentrations often unrealistically high. Effects most likely to be realized at low concentrations of petroleum contamination seem to be those involving disruption of chemoreception ("smell") and consequent behavioral alterations. Crude oil in dissolved concentrations of the order of 20 to 40 p.p.b. caused modifications of feeding behavior in lobsters (113). Concentrations of dissolved petroleum hydrocarbons as low as 1 to 4 p.p.b. interfered with sensing and location of food by marine snails (114). Thus it appears that quite low level contamination by oil might cause "odors" that affect the senses of marine organisms in such a manner as to alter their behavior.

Of great concern is the danger of cancer being induced in man and animals from ingesting or contacting oil. Crude oils and refined products contain known carcinogenic compounds, mainly polycyclic aromatics, in small quantities (2). Concern has been expressed for the human health hazards involved in eating seafood contaminated with oil and thus with carcinogenic material (2,115). Although the amount of carcinogens potentially consumed in this manner (115) seems small when compared to other sources such as green vegetables and roast meat (116), the prevailing philosophy favors a conservative view that there is no lower threshold of carcinogens in the body and that any increases should be avoided (116).

Similarly no conclusive evidence exists for carcinogenic effects on marine organisms. However, circumstantial evidence suggests that petroleum hydrocarbons may be involved in initiating tumors in fishes (4) and clams (117,118).

Chapter Six

Evaluation

WHAT IS KNOWN

Amid the confusion generated by disparate scientific reports, overblown statements by conservationists, industrial propagandists, and the news media, and the intense emotionalism of the oil pollution issue, it remains necessary to summarize our knowledge of the ecological effects of oil spills and place them in perspective with other energy-environment problems. A good start is to summarize the detrimental effects we know spilled oil *can* have.

Oil can kill marine life directly through:

(1) Coating and asphyxiation (example: barnacles and other intertidal life);
(2) Poisoning through direct contact or ingestion (examples: ingestion of oil by preening birds, contact poisoning of vascular plants);
(3) Exposure to water-soluble toxic petroleum components (example: Subtidal fishes and invertebrates at Midway Island, the *Tampico Maru* spill, and West Falmouth);
(4) Destruction of more sensitive juvenile forms (example: fish eggs and larvae); and
(5) Disruption of body insulation of warm blooded animals (example: diving birds).

Oil may also have harmful indirect effects, including:

(1) Destruction of food sources;
(2) Synergistic effects that reduce resistance to other stresses;
(3) Incorporation of carcinogenic and potentially mutagenic chemicals;
(4) Reduction of reproductive success; and
(5) Disruption of chemical clues essential to survival, reproduction, or feeding.

There is no serious opposition to the above observations, at least not in the hypothetical sense. Where differences of opinion arise is over which of these effects will actually come to bear following a given oil spill. From the many published scientific observations of oil spills one might conclude that acute ecological damage was in most cases "light" to "moderate" (Milford Haven, Santa Barbara, San Francisco, and Chedabucto Bay spills), but in a few cases damage was "severe" (*Tampico Maru*, West Falmouth and, because of the misuse of toxic dispersants, *Torrey Canyon*).

Of course, spilled oil affects different organisms differently. Oil pollution has consistently done direct and significant damage to diving sea birds to the point of threatening the local survival of several species. This in itself is a most serious consequence of oil spills. Most observations have been confined to or have concentrated on visible intertidal organisms, which while certainly susceptible to heavy dosages of oil, are notoriously hardy marine organisms. It must be kept firmly in mind when judging the severity of oil spills as reported in the literature that the literature usually ignores the less easily visible, less tolerant subtidal organisms (119).

The long-term effects of oil spills are too little known to make definitive conclusions. A British pollution biologist, R.B. Clark (87), concluded that "even catastrophic coastal pollution, combined with the grossest misuse of toxic dispersants, appears to do no irreparable harm, although recovery of the flora and fauna may take some years." However, his statement applied to those situations where recovery is possible, i.e., where pollution is not chronic, and to temperate and subtropical waters. On the other hand, follow-up studies of the fuel oil spills at West Falmouth (19,63,64) and in Baja California (10) indicate that recovery can be slow and pollution relatively persistent from a single spill. The observable long-term effects of a single oil spill seem to range from less than one year to more than a decade after the initial accident.

Chronic pollution, via outfall, land runoff, or ship discharge may have consequences of much longer duration. Chronically polluted waters near oil terminals and refineries may be severely perturbed environments, and exclusion of normal plant and animal species may continue for as long as pollution continues. The effects of large, chronic inputs into the marine environment from sewage, runoff, and tanker operation have not been investigated but clearly warrant concern. It is especially important to note here that chronic discharges of oil are not evenly spread throughout the world's oceans, but almost all occur in coastal waters, the most productive part of the sea.

FACTORS INFLUENCING SEVERITY OF ECOLOGICAL EFFECTS

What, then, determines the severity of an oil spill? A large number of factors are no doubt important, but a few stand out:

(1) The dosage of oil an environment receives;
(2) The physical and chemical nature of the oil spilled, including the effects of weathering;
(3) The location of the spill;
(4) The time of year of the spill;
(5) The prevailing weather conditions; and
(6) The techniques used to clean up the spill (18).

Spills in which the oil was concentrated in a small area (e.g. *Tampico Maru* and West Falmouth spills) have been more damaging than those in more open waters. However, dosage is difficult to estimate and thus difficult to relate to ecological effect. Dosage by soluble, floating, and sunken oil are independent problems, and only floating dosage has been estimated with any success (120).

The relative importance of the type of oil spilled is controversial. A much publicized difference of opinion has developed about the Santa Barbara oil spill between Dr. Dale Straughan of the University of Southern California and Dr. Max Blumer of the Woods Hole Oceanographic Institution. This controversy centers in part on the relative importance of the type of oil spilled and in part on methodological differences. Straughan led the USC team studying the effects of the blowout and observed what she thought was a surprisingly small effect on marine life. Comparing this with the more devastating spills from the *Tampico Maru* and *Florida*, she concluded that the difference in the chemical and physical properties of the oils were the factors of primary importance (18). Santa Barbara crude tended to float, was relatively insoluble in water, and was low in aromatic hydrocarbons, whereas the Number 2 fuel oil spilled at West Falmouth dispersed rapidly in water and was high in toxic aromatics. Thus, the effects of the Santa Barbara spill were basically the smothering of intertidal animals and the fouling of birds, while the West Falmouth oil had severe toxic effects on a wide variety of organisms. Blumer (14) attributes the discrepancy in findings mainly to the disparate methods of investigation. He notes that fuel oil is a part of crude oil; thus, the toxic hydrocarbons found in fuel oil are also found in crude and should certainly be toxic. Furthermore, the investigations at West Falmouth focused on small benthic animals in contact with high concentrations of oil in the sediments, and on the chemical analyses of contamination in organisms and sediments. Santa Barbara studies concentrated on resistant intertidal organisms and large migratory animals, and very little chemical investigation was done. In short, Blumer concludes that the Santa Barbara investigations were not designed to uncover the effects found at West Falmouth.

Although Blumer's criticisms, particularly those on the lack of adequate benthic and chemical studies, are well taken, it is difficult to envisage how extensive kills like those at West Falmouth could have gone unnoticed in the Santa Barbara Channel, which was being studied by investigators from many organizations. The deficiencies of many oil pollution studies, including those at Santa Barbara, are regrettable. But one must conclude from studies around the world that the ecological effects depend greatly on the nature of the oil spilled. On the other hand, it does not seem reasonable to conclude, as have some oil company spokesmen and consultants (121,122), that the West Falmouth spill was atypical in effect due to the type of oil, confinement of the spill, and weather conditions. Studies of the West Falmouth spill suggest that the same kinds of effects, even if not of equivalent magnitude, could be experienced following a spill of any type of oil, even crude oil (2,119).

The geography of the spill is of obvious importance. We have already speculated on spills in polar and tropical regions and in estuaries. Also the biotic nature of habitats varies greatly within temperate zones; for example, the east coast of the United States is geologically and ecologically quite different from the west coast, and the Louisiana coastal environment is quite unlike that of Maine.

The time of the year the spill occurs may also be critical. Had the Santa Barbara spill happened earlier, nursing pups of sea lions and elephant seals may have succumbed after ingesting oil coating their mothers teats, and sea bird populations would have been greater. Also a spill could endanger the success of a seasonal reproductive period.

Meteorological and oceanographic conditions influence the effects of an oil spill. Floating oil responds primarily to winds and currents and may be blown inshore or offshore. Floating oil from the Santa Barbara spill met sediment-laden plumes of low salinity runoff waters generated by atypically heavy rains (69). Oil mixed with the sediments, sank, and settled on the bottom rather than on the shore. Similarly, at West Falmouth onshore winds churning oil with sediment deposited oil on the bottom, killing many bottom animals (14).

An improper clean-up strategy can worsen the effect of oil pollution. The *Torrey Canyon* spill taught us the lesson of rampant misuse of toxic dispersants. The sinking of *Torrey Canyon* oil off Brittany shifted its impact from the intertidal to the subtidal benthic environment, where more areas are affected and the oil may persist.

SHORTCOMINGS OF RESEARCH ON OIL POLLUTION

Although the literature on effects of oil in the marine environment is of considerable bulk, on balance ecologists have failed to come to grips with the biological effects of oil. Certainly, this may also be said of other seaborne pollutants we know very little about. Yet we know even less about oil than about heavy metals and pesticides, for example. Although there is undoubtedly a complex array of cultural, economic, and political explanations for this generally poor state of knowledge, we offer only a critique of research on ecological effects.

Field observations have been and should be of primary importance in the assessment of the effects of oil pollution. In the past these observations have often been casual and non-quantitative, which precludes discerning less-than-catastrophic ecological effects, fails to provide sufficient information to allow other scientists to judge these effects for themselves, and thus may leave in question the validity of the observations. Unbiased sampling, quantification, replication, and statistical treatment should be employed in ecological studies in the field. Hard facts rather than casual observations will be more and more required, especially if such data are to be used as legal evidence.

Most field studies have been limited to or have concentrated on intertidal organisms. This is because of (1) the long-held beliefs that oil always floats, (2) the visibly heavy dosage a shore may receive after a spill, (3) a long tradition of study of the intertidal environment, and (4) the fact that this environment is much easier to study than subtidal ones (special equipment and vessels are not required, and one can readily observe the intertidal inhabitants). Many researchers have labored under the mistaken impression that the effects of oil on organisms below the water's surface is negligible. As has been shown in this report, this may not be the case, and oil may affect subtidal organisms, particularly those that live on or in the bottom–another place where oil may concentrate. More research should be done on the effects of oil spills and chronic oil pollution on subtidal organisms.

Oil pollution has almost always been measured by its lethal effect on adult organisms. Very little has been done on the effects experienced by larvae or juveniles or on the sublethal effects on reproductive success. The little information available indicates that larvae and juveniles are usually more sensitive to pollutants than are adults, and that oil may inhibit the production of gametes or otherwise reduce the chances of reproductive success. It is the reproductive and recruitment links of life cycles that are often most crucial and that determine whether and how an organism can exploit an environment.

Until recently ecologists have overlooked or failed to address the problem of oil uptake and contamination in marine organisms. The study of contamination necessarily involves chemical analyses beyond the experience of most ecologists. On the other hand, interpretations of petroleum contamination by chemists have sometimes shown a lack of understanding of biological processes and effects.

Research on the biodegradation of oil has failed to adequately describe its processes in nature. While numerous studies have demonstrated the rapid breakdown of oil under ideal laboratory conditions, rates of degradation in the marine environment, particularly of oil in sediments and in the open sea, remain unknown.

Ecosystem recovery following oiling has usually not been adequately studied. There are practical impediments to the long-term research and funding commitments required to investigate a five to ten-year period of recovery; however, information on long-term recovery of ecosystems is essential to any assessment of environmental damage by oil pollution.

Laboratory studies have also had serious shortcomings (123), and one shortcoming is inherent in all laboratory ecological studies: the limitation of extrapolating laboratory experimental results to the real world of nature. The few species selected for study cannot be considered representative of the response of the whole community. Typically, more tolerant and hardy species are chosen because they "do well in the lab." Often tests are made only on adults, although it is generally the larval and juvenile forms that are most sensitive. Many bioassay studies have lacked statistical rigor and are therefore of questionable validity. Compounding these basic limitations is the special difficulty in approximating the dosage of petroleum that is largely insoluble; this is less of a problem for bioassays of soluble dispersants. Most often the concentration of petroleum hydrocarbons in the water tested has not even been known. The most useful roles of laboratory bioassays seem to be in assessing relative toxicities of oils or dispersants, explaining phenomena observed in the field, and investigating toxicological mechanisms.

Most laboratory studies have been limited to acute (of two to four days duration) lethal effects. Chronic and sublethal effects have not been well studied. However, it is in these areas where laboratory studies may have their greatest value, considering the difficulty in controlling variables and running long-term experiments in nature.

RESEARCH NEEDS

Now that the shortcomings of research on the effects of oil pollution have been examined, some positive suggestions as to the direction of future research seem in order. The following broad areas of research are considered priorities:

1. More detailed field investigations of oil spills are needed to more fully comprehend the ecological effects of such incidents. Despite the voluminous reports on the effects of *Torrey Canyon*, Santa Barbara, and other spills, our basic understanding of such effects does not permit general agreement on their severity, nor does it permit reasonable predictions. These field studies should be multidisciplinary (including at least biological and chemical investigations), not limited to the intertidal biota, carefully designed, meticulously carried out, statistically rigorous, and of several (three or more) years duration. It is unrealistic to expect the tooling-up required for this task to be accomplished immediately subsequent to an accidental spill—at least at all but a few research institutions. For this reason, spill simulation and field experimentation are preferred approaches to the study

of immediate ecological effects and ecosystem recovery. In this way, the type and dosage of oil, as well as several other factors, can be controlled. This does not mean that accidental spills should not be investigated–even on a shoestring basis–for considerable knowledge can be derived from these studies, and they are the ultimate testing grounds for experimental conclusions.

2. The effects of chronic or multiple pollution by oil need increased attention. Detailed ecological investigations should be undertaken in ecosystems receiving refinery and sewage hydrocarbon wastes and in oil ports. Estuarine ecosystems already under stress deserve special attention because of their importance to coastal productivity. This problem seems especially pertinent in light of increasing oil imports and the demand for new coastal refinery capacity.
3. Experimental research is needed on the sublethal effects of oil on marine organisms, including effects on photosynthesis, metabolism, reproduction, behavior, and chemical reception.
4. The uptake, retention, and release of petroleum hydrocarbons by marine organisms should be further studied. Background levels of "natural" and petroleum-derived hydrocarbons need quantification and qualification. Mechanisms of uptake, metabolic fate, and release are ill-studied. Knowledge of the persistence of various hydrocarbons in organisms and the non-living parts of the biosphere is required.
5. The threat of oil pollution to human health needs further study and evaluation. Carcinogenic contamination of fishery organisms deserves special attention.
6. Finally, much of the lack of understanding about the effects of oil pollution stems from our lack of understanding of marine ecosystems. More baseline information is needed against which the effects of pollutants of all kinds may be measured.

SUPPORT OF RESEARCH

The direction as well as the volume of oil pollution research depends directly on the sources and levels of financial support for that research. The interplay of public (governmental) and private (industrial) support reflects various political, economic, and public opinion pressures, sometimes acting at variance. In no other area of marine pollution research, except perhaps for that related to thermal additions by power plants, is industrial involvement so heavy. At least in part, this explains why there are more conflicting opinions among researchers in this area than in any other. Although an analysis of the administration of oil pollution research is peripheral to the scope of this report, a brief discussion of the sources of support and their effect on objective research seems in order in the context of recommendations for further research.

The Federal government has supported ecological research on oil pollution through a number of agencies, notably the Environmental Protection Agency (EPA), the Coast Guard, the Navy, the National Oceanic and Atmospheric Administration (NOAA), and the National Science Foundation (NSF). This support has been for inhouse research as well as through extramural grants and contracts. Most support has been for highly mission-oriented research, for example for the study of a specific oil spill or for the investigation of accelerating biodegradation. Most of the Coast Guard and Navy research has been on the fate and control of spilled oil, and it is the EPA that has been the principal sponsor of research on ecological effects. Ongoing EPA research is centered at the Edison Water Quality Research Laboratory and is "fate" oriented. EPA's extramural funds for ecological effects studies are currently about $400,000 per annum, which allows the funding of only three major research efforts, one of which accounts for over 85 percent of the funds. NSF's International Decade of Ocean Exploration program is sponsoring research on problems related to chronic pollution of the sea, including an investigation of the effects of oil on the *Sargassum* community and various studies on sublethal effects. Considered in light of the magnitude of the problem and the level of support of other water pollution programs, one must conclude that federal funding of research on the effects of oil pollution is grossly deficient. This support needs to be increased, and EPA is the logical agency to administer a broadened research program.

State governments have and are continuing to conduct or sponsor research on oil pollution. Particularly notable are the states of Maine and Washington, which are experiencing extensive oil terminal and refinery development.

The petroleum industry has sponsored considerable ecological research since the 1940s. This support may emanate from individual companies, regional consortia of companies, or the national industry organization: the American Petroleum Institute (API). Most corporate support is via consultantships, and the results of research are often proprietary. The Western Gas and Oil Association is funding the continuing University of Southern California Santa Barbara studies as well as others in Alaskan waters. A group of Gulf Coast oil producers is funding a large number of ecological studies at a number of Southern universities through the Gulf Universities Research Corporation. Finally, the API sponsors a number of ecological research efforts (124) including biodegradation studies, research on the acute and sublethal (chronic) effects of oil, and observations of the recovery of the West Falmouth spill area. It is very important to note that, over all, industry is a larger sponsor of oil pollution research than is the public sector.

Although some good research has been sponsored by the petroleum industry, particularly the API, it should be noted that much of it was commissioned to serve the interests of the industry. In some cases industry-sponsored research has been virtually the only source of information on ecological effects on an oil spill. Frequently this information is kept confidential pending litigation in the courts. Such has been the status of the extensive research on the effects of the Louisiana blowouts of recent years. This does not allow adequate opportunity for critical evaluation by the scientific community. Some scientists have suggested that such unreviewed results and interpretation be inadmissible as legal evidence (15). Otherwise, they contend, a serious imbalance in the legal adversary system may occur. Certainly, oil companies have a right to engage consultants to assemble convincing evidence in their defense, and thus the blame for any information imbalance belongs in the public sector.

Why, then, is not more public-sponsored research being done on the important problems of oil pollution? The most obvious reason is lack of financial support by government. This is probably because oil pollution research is relatively new and has not developed the large scientific and bureaucratic constituency as have other fields of pollution research such as pesticide, thermal pollution, and eutrophication research. One also wonders about the role of political and economic pressures in the continuing paucity of governmental research funds. Oil pollution research, if properly conducted, is expensive because the multidisciplinary approach required involves costly field and laboratory equipment and much manpower. Research sponsors, seeking quicker and more tangible results from their investments, have often preferred to fund symposia to look into the problem rather than underwrite lengthy and expensive research.

However, part of the blame must rest with the scientific community as well. Marine ecologists, at least those in the U.S., have responded lethargically to the threats of oil pollution. Their reluctance to address the problem is at least in part due to the great length of time required for research on the effects of oil, which precludes immediate rewards of publishable results.

Chapter Seven

Conclusions

The ecological effects of oil pollution on the marine environment will be an important consideration in energy policy decisions in the future. Public pressures and legal mandates, such as the National Environmental Policy Act and the Federal Water Pollution Control Act, will insure this. Changes in policies governing oil imports will affect the possibilities of accidental spills. International agreements concerning intentional shipping discharges will be formulated. Decisions will be made on where to allow offshore oil exploration and production, and on the types of pollution prevention technology required in these production fields. Superports will be planned, as will coastal refineries.

At present, assessment of the environmental impact of such developments must be made in considerable ignorance and uncertainty because of large knowledge gaps and conflicting opinions. Because so many serious questions remain unanswered, and because of the alarming implications of some of the information available, we recommend great caution in making policy decisions involving oil and the marine environment. Given the diverse and often equivocal evaluations offered by the scientific community, it falls to society to decide what level of confidence to place in available information concerning the consequences of oil pollution of the marine environment. Do we assume a pollutant is "innocent" until proven "guilty," as we have often done in the past? Or do we assume it is "guilty" until proven "innocent," as we currently do with drugs? Or shall we scrupulously avoid making assumptions and seek the full range of scientific information needed to arrive at well-considered judgments?

The only remedy for our uncomfortable ignorance is more and better research into the problem—especially into the more neglected aspects, such as chronic pollution and sublethal effects. It is interesting to note that, while not implying that oil pollution is a necessarily equivalent problem, it took many years of research on persistent pesticides—much more time than has been spent on the effects of oil—to affect policy decisions resulting in the control of

DDT alone. We are still a long way from this level of commitment to understanding the effects of those complex chemical mixtures we know as oil.

Bibliography

1. Study of Critical Environmental Problems, *Man's Impact on the Global Environment: Assessment and Recommendations for Action:* MIT Press, Cambridge, 319 pp., 1970.
2. Blumer, M. *Scientific Aspects of the Oil Spill Problem. Environmental Affairs*, 1, pp. 54-73, 1971.
3. International Maritime Consultative Organization, *The Environmental and Financial Consequences of Oil Pollution from Ships.* Preparations for International Marine Pollution Conference 1973. United Kingdom Programmes Analysis Unit, Chilton, Didcot, Berks, 1973.
4. Nelson-Smith, A. *Oil Pollution and Marine Ecology.* Elek Science, London. 260 pp., 1972.
5. Clark, R.B. *Oil Pollution and its Biological Consequences. A Review of Current Scientific Literature.* Prepared for Great Barrier Reef Petroleum Drilling Royal Commissions, 111 p., 1971.
6. Galtsoff, P.S., Prytherch, H.F., Smith, R.O., and Koehring, V. *Effects of Crude Oil Pollution on Oysters in Louisiana Waters.* Bulletin of the Bureau of Fisheries, Washington, D.C. 48, pp. 143-210, 1935.
7. Lunz, G.R. *The Effect of Bleedwater and of Water Extracts of Crude Oil on the Pumping Rate of Oysters.* Texas A & M Research Foundation, 1950. (Project Nine–unpublished report.)
8. Mackin, J.G. *A Report on Three Experiments to Study the Effect of Oil Bleedwater on Oysters Under Aquarium Conditions.* Texas A & M Research Foundation, 1950. (Project Nine–unpublished report.)
9. Menzel, R.W. *Report on Two Cases of "Oily Tasting" Oysters at Bay Ste. Elaine Oilfield.* Texas A & M Research Foundation, 1948. (Project Nine–unpublished report.)
10. North, W.J., Neushul, M., Jr., and Clendenning, K.A. *Successive Biological Changes Observed in a Marine Cove Exposed to a Large Spillage of Oil. Symposium Commission internationale exploration scientifique Mer Mediterranee, Monaco, 1964*, 335-354, 1965.

11. Cowell, E.B., editor. *Proceedings of the Symposium on the Ecological Effects of Oil Pollution of Littoral Communities, London, 30 November-1 December, 1970.* London, Institute of Petroleum, 1971.
12. Smith, J.E., editor. *'Torrey Canyon' Pollution and Marine Life: A Report by the Plymouth Laboratory.* xiv, 196 p. Cambridge University Press, 1968.
13. Allan Hancock Foundation. *Biological and Oceanographical Survey of the Santa Barbara Channel Oil Spill 1969-1970.* 2 vols. Allan Hancock Foundation, University of Southern California, 1971. Vol. 1. *Biology and Bacteriology*; compiled by D. Straughan. 426 p. Vol. 2. *Physical, Chemical and Geological Studies*; general editor, R.L. Kolpack. 477 p.
14. Blumer, M., Sanders, H.L., Grassle, J.F., and Hampson, G.R. *A Small Oil Spill. Environment*, 13, (2), pp. 1-12, 1971.
15. Neushul, M., Jr. *Effects of Pollution on Populations of Intertidal and Subtidal Organisms.* Paper presented at Santa Barbara oil symposium, *Offshore Petroleum Production–An Environmental Inquiry*, Santa Barbara, California, 16-18 December, 1970.
16. Goldacre, R.J. *The Effects of Detergents and Oils on the Cell Membrane. Field Studies*, 2, (suppl.) pp. 131-137, 1968.
17. Falk, H.L., Kotin, P., and Hehler, A. *Polycyclic Hydrocarbons as Carcinogens for Man. Archives of Environmental Health*, 8, pp. 721-730, 1964.
18. Straughan, D. *Factors Causing Environmental Changes After an Oil Spill. Journal of Petroleum Technology*, March 1972, pp. 250-254, 1972.
19. Blumer, M., and Sass, J. *The West Falmouth Oil Spill*, data available in November, 1971. II. *Chemistry.* Technical Report of the Woods Hole Oceanographic Institution No. 72-19, 60 pp., 1972.
20. Zobell, C.E. *Action of Microorganisms on Hydrocarbons. Bacteriology Review*, 10, pp. 1-49, 1946.
21. _______. *Microbial Modification of Crude Oil in the Sea.* In: American Petroleum Institute. *Proceedings of a Joint Conference on Prevention and Control of Oil Spills*, New York, December 15-17, 1969, pp. 317-336. New York, American Petroleum Institute, 1970.
22. Clark, R.B. *Oil Pollution and the Conservation of Seabirds.* Proceedings of the International Conference on Oil Pollution. *Sea*, Rome, pp. 76-112, 1968.
23. Straughan, D. *Oil Pollution and Sea Birds.* In: Allan Hancock Foundation. *Biological and Oceanographical Survey of the Santa Barbara Channel Oil Spill 1969-1970.* Vol. 1, compiled by D. Straughan, pp. 307-312. Allan Hancock Foundation, University of Southern California, 1971.
24. Chan, G.L. *A Study of the Effects of the San Francisco Oil Spill on Marine Organisms.* In: American Petroleum Institute. *Proceedings of a Joint Conference on Prevention and Control of Oil Spills, Washington, D.C., March 13-15, 1973.*, pp. 739-781. Washington, D.C., American Petroleum Institute, 1973.
25. Connell, J.H. *Submission to the Royal Commission on Oil Exploitation on the Great Barrier Reef*, May 1971.
26. Westfall, A. *Jackass Penguins. Marine Pollution Bulletin*, 14, pp. 2-7, 1969.
27. Clark, R.B. Personal communications, 1973.

28. Brownell, R.L., Jr., and LeBoeuf, B.J. *California Sea Lion Mortality: Natural or Artifact?* In: Allan Hancock Foundation. *Biological and Oceanographical Survey of the Santa Barbara Channel Oil Spill 1969-1970.* Vol. 1, compiled by D. Straughan, pp. 255-276. Allan Hancock Foundation, University of Southern California, 1971.
29. LeBoeuf, B. *Oil Contamination and Elephant Seal Mortality: A "Negative" Finding.* In: Allan Hancock Foundation. *Biological and Oceanographical Survey of the Santa Barbara Channel Oil Spill 1969-1970.* Vol. 1, compiled by D. Straughan, pp. 277-285. Allan Hancock Foundation, University of Southern California, 1971.
30. St. Amant, L.S. *Biological Effects of Petroleum Exploration and Production in Coastal Louisiana.* In: Holmes, R.W., and DeWitt, F.A., editors. *Santa Barbara Oil Symposium, Santa Barbara, December 16-18, 1970*, University of California, 1970.
31. Tendron, G. *Contamination of Marine Flora and Fauna by Oil, and the Biological Consequences of the 'Torrey Canyon' Accident.* Proceedings of the International Conference on Oil Pollution. *Sea*, Rome, 1968, pp. 114-121, 1969.
32. Diaz-Piferrer, M. *The Effects of an Oil Spill on the Shore of Guanica, Puerto Rico.* (Abstract) *Association of Island Marine Laboratories*, 4th Meeting, Curacao, 12-13, 1962.
33. Gooding, R.M. *Oil Pollution on Wake Island from the Tanker* R.C. Stoner. National Oceanic and Atmospheric Administration, National Marine Fisheries Service, *Special Report–Fisheries*, No. 636, 12 p., 1971.
34. Ebeling, A.W., DeWitt, F.A., Werner, W., and Cailliet, G.M. *Santa Barbara Oil Spill: Fishes.* In: Holmes, R.W., and DeWitt, F.A., editors. *Santa Barbara Oil Symposium, Santa Barbara, December 16-18, 1970.* University of California, 1970.
35. Sherman, K., Colton, J.B., Knapp, F.R., and Dryfoos, R.L. *Fish Larvae Found in Environment Contaminated with Oil and Plastic.* National Marine Fisheries Service, MARMAP Red Flag Report No. 1, 1973.
36. Offshore Oil Task Group, Massachusetts Institute of Technology. *The Georges Bank Petroleum Study, Volume II.* Massachusetts Institute of Technology, Cambridge, 311 pp., 1973.
37. Vale, G.H., Sidhu, G.S., Montgomery, W.A., and Johnson, A.R. *Studies of a Kerosene-like Taint in Mullet* (Mugil cephalus). *Journal of the Science of Food and Agriculture*, 21, pp. 429-432, 1970.
38. Shipton, J., Last, J.H., Murray, K.E., and Vale, G.L. *Studies on a Kerosene-like Taint in Mullet* (Mugil cephalus). *Journal of the Science of Food and Agriculture*, 21, pp. 433-436, 1970.
39. Straughan, D. *Oil Pollution and Fisheries in the Santa Barbara Channel.* In: Allan Hancock Foundation. *Biological and Oceanographical Survey of the Santa Barbara Channel Oil Spill 1969-1970.* Vol. 1, compiled by D. Straughan, pp. 245-254. Allan Hancock Foundation, University of Southern California, 1971.
40. Mead, W.J., and Sorensen, P.E. *The Economic Cost of the Santa Barbara Oil Spill.* In: Holmes, R.W., and DeWitt, F.A., editors. *Santa Barbara Oil*

Symposium, Santa Barbara, December 16-18, 1970, University of California, 1970.

41. Hawkes, A.L. *A Review of the Nature and Extent of Damage Caused by Oil Pollution at Sea. Transactions of the North American Wildlife Conference*, 26, pp. 343-355, 1961.
42. Blumer, M., Souza, G., and Sass, J. *Hydrocarbon Pollution of Edible Shellfish by an Oil Spill. Marine Biology*, 5, pp. 195-202, 1970.
43. Ehrhardt, M. *Petroleum Hydrocarbons in Oysters from Galveston Bay. Environmental Pollution*, 3, pp. 257-272, 1972.
44. Lee, R.F., Sauerheber, R., and Benson, A.A. *Petroleum Hydrocarbons: Uptake and Discharge by the Marine Mussel,* Mytilus edulis. *Science*, N.Y., 177, pp. 344-346, 1972.
45. Anderson, J. Personal communication, 1973.
46. Lee, R.F., Sauerheber, R., and Dobbs, G.H. *Uptake, Metabolism and Discharge of Polycyclic Aromatic Hydrocarbons by Marine Fish. Marine Biology*, 17, pp. 201-208, 1972.
47. Spears, R.W. *An Evaluation of the Effects of Oil, Oil Field Brine, and Oil Removing Compounds.* In: American Institute of Mining, Metallurgical and Petroleum Engineers. *AIME Environmental Quality Conference, Washington, D.C., June 7-9, 1971,* pp. 199-216. American Institute of Mining, Metallurgical, and Petroleum Engineers.
48. Oguri, M., and Kanter, R. *Primary Productivity in the Santa Barbara Channel.* In: Allan Hancock Foundation. *Biological and Oceanographical Survey of the Santa Barbara Channel Oil Spill 1969-1970.* Vol. 1. compiled by D. Straughan, pp. 17-48. Allan Hancock Foundation, University of Southern California, 1971.
49. McGinnis, D.R. *Observations on the Zooplankton of the Eastern Santa Barbara Channel from May 1969 to March 1970.* In: Allan Hancock Foundation. *Biological and Oceanographical Survey of the Santa Barbara Channel Oil Spill 1969-1970.* Vol. 1, compiled by D. Straughan, pp. 49-59. Allan Hancock Foundation, University of Southern California, 1971.
50. Galtsoff, P.S., Prytherch, H.F., Smith, R.O. and Koehring, V. *Effects of Crude Oil Pollution on Oysters in Louisiana Waters. Bulletin of the Bureau of Fisheries*, Washington, D.C., 48, pp. 143-210, 1935.
51. Mironov, O.G. *Hydrocarbon Pollution of the Sea and its Influence on Marine Organisms. Helgoländische wissenschaftliche Meeresunters uchungen,* 17, pp. 335-339, 1968.
52. Aubert, M., Charra, R. and Malara, G. *Etude de la toxicite de produits chimiques vis-à-vis de la chaine biologique marine. Révue internationale d'oceanographie medicale*, 13/14, 45-72, 1969.
53. Kauss, P., Hutchinson, T.C., Soto, C., Hellebust, J., and Griffiths, M. *The Toxicity of Crude Oil and its Components to Freshwater Algae.* In: American Petroleum Institute. *Proceedings of a Joint Conference on Prevention and Control of Oil Spills, Washington, D.C., March 13-15, 1973,* pp. 703-714. Washington, American Petroleum Institute, 1973.
54. Strand, J.A., Templeton, W.L., Lichatowich, J.A., and Apts, C.W. *Development of Toxicity Test Procedures for Marine Phytoplankton*. In: American

Petroleum Institute. *Proceedings of a Joint Conference on Prevention and Control of Oil Spills, Washington, D.C., June 15-17, 1971*, pp. 279-286. Washington, American Petroleum Institute, 1971.

55. Nuzzi, R. *Effects of Water Soluble Extracts of Oil on Phytoplankton.* In: American Petroleum Institute. *Proceedings of a Joint Conference on Pollution and Control of Oil Spills, Washington, D.C., March 13-15, 1973*, pp. 809-813. Washington, American Petroleum Institute, 1973.
56. Gordon, D.C., and Prouse, N.J. *The Effects of Three Oils on Marine Phytoplankton Photosynthesis. Marine Biology*, 22, pp. 329-333, 1973.
57. Sponner, M.F. *Effects of Oil and Emulsifiers on Marine Life.* In: Hepple, P., editor. *Water Pollution by Oil: Proceedings of a Seminar Held at Aviemore . . . 4-8 May 1970,* pp. 375-376. London, Institute of Petroleum, 1971.
58. Kuhnhold, W.W. *Effect of Water Soluble Substances of Crude Oil on Eggs and Larvae of Cod and Herring,* 15 p. Copenhagen, International Council for the Exploration of the Sea, Fisheries Improvement Committee, 1969. (CM 1969/E 17).
59. Burns, K.A. and Teal, J.M. *Hydrocarbons in the pelagic* Sargassum *community. Deep-Sea Research* 20, pp. 207-211, 1973.
60. Ottway, S. *The Comparative Toxicities of Crude Oils.* In: Cowell, E.B., editor. *Proceedings of the Symposium on the Ecological Effects of Oil Pollution on Littoral Communities, London, 30 November-1 December, 1970.* London, Institute of Petroleum, 1971.
61. Nicholson, N.L., and Climberg, R.L. *The Santa Barbara Oil Spills of 1969: A Post-Spill Survey of the Rocky Intertidal.* In: Allan Hancock Foundation. *Biological and Oceanographical Survey of the Santa Barbara Channel Oil Spill 1969-1970.* Vol. 1, compiled by D. Straughan, pp. 325-399. Allan Hancock Foundation, University of Southern California, 1971.
62. Foster, M., Neushul, M., and Zingmark, R. *The Santa Barbara Oil Spill: Part 2–Initial Effects on Intertidal and Kelp Bed Organisms. Environmental Pollution*, 2, pp. 115-134, 1971.
63. Burns, K.A., and Teal, J.M. *Hydrocarbon Incorporation into the Salt Marsh Ecosystem from the West Falmouth Oil Spill.* Technical Report of the Woods Hole Oceanographic Institution, No. 71-69, 24 p., 1971.
64. Sanders, H.L., Grassle, J.F., and Hampson, G.R. *The West Falmouth Oil Spill.* I. *Biology*. Technical Report of the Woods Hole Oceanographic Institution, No. 72-20, 23 p., 1972.
65. Thomas, M.L.H. *Effects of Bunker C Oil on Intertidal and Lagoonal Biota in Chedabucto Bay, Nova Scotia. Journal of the Fisheries Research* Bd. Canada, 30, pp. 83-90, 1973.
66. Straughan, D. Personal communication, 1973.
67. Crapp, G.B. *Chronic Oil Pollution.* In: Cowell, E.B., editor. *Proceedings of the Symposium on the Ecological Effects of Oil Pollution on Littoral Communities, London, 30 November-1 December, 1970.* London, Institute of Petroleum, 1971.
68. Connell, J.H. Personal communication, 1973.
69. Drake, D.E., Fleischer, P., and Kolpack, R.L. *Transport and Deposition of*

Flood Sediment, Santa Barbara Channel, California. In: Allan Hancock Foundation. *Biological and Oceanographical Survey of the Santa Barbara Channel Oil Spill 1969-1970.* Vol. 2, general editor, R.L. Kolpack, pp. 181-217. Allan Hancock Foundation, University of Southern California, 1971.

70. Kolpack, R.L., Mattson, J.S., Mark, H.G., Jr., and Yu, T.-C. *Hydrocarbon Content of Santa Barbara Channel Sediments.* In: Allan Hancock Foundation. *Biological and Oceanographical Survey of the Santa Barbara Channel Oil Spill 1969-1970.* Vol. 2, general editor, R.L. Kolpack, pp. 276-295. Allan Hancock Foundation, University of Southern California, 1971.
71. Scarratt, D.J., Sprague, J.B., Wilder, D.G., Zitko, V., and Anderson, J.M. *Some Biological and Chemical Investigations of a Major Winter Oil Spill on the Canadian East Coast.* 8 p. International Council for the Exploration of the Sea, Fisheries Improvement Committee, 1970 (CM 1970/E 14).
72. Fauchald, K. *The Benthic Fauna in the Santa Barbara Channel Following the January 1969 Oil Spill.* In: Allan Hancock Foundation. *Biological and Oceanographical Survey of the Santa Barbara Channel Oil Spill 1969-1970.* Vol. 1, compiled by D. Straughan, pp. 61-116. Allan Hancock Foundation, University of Southern California, 1971.
73. Kolpack, R.L. Personal communication, 1973.
74. Hampson, G.R., and Sanders, H.L. Local oil spill. *Oceanus*, 15(2), pp. 8-11, 1969.
75. Sanders, H.L. Personal communication, 1973.
76. Scarratt, D.J., and Zitko, V. *Bunker C Oil in Sediments and Benthic Animals from Shallow Depths in Chedabucto Bay, N.S. Journal of the Fisheries Research Board of Canada*, 29, pp. 1347-1350.
77. Mackin, J.G. *Report on a Study of the Effect of Application of Crude Petroleum on Saltgrass* Distychlis spicata. Texas A & M Research Foundation, 1950.
78. _______. *A Comparison of the Effect of Application of Crude Petroleum to Marsh Plants and to Oysters.* Texas A & M Research Foundation, 1950. (Project Nine—unpublished report.)
79. Stebbings, R.E. *Recovery of Salt Marsh in Brittany Sixteen Months after Heavy Pollution by Oil. Environmental Pollution*, 1, pp. 163-167, 1970.
80. Cowell, E.B. *The Effects of Oil Pollution on Salt Marsh Communities in Pembrokeshire and Cornwall. Journal of Applied Ecology*, 6, pp. 133-142, 1969.
81. Baker, J.M. *Successive Spillages.* In: Cowell, E.B., editor. *Proceedings of the Symposium on the Ecological Effects of Oil Pollution on Littoral Communities, London, 30 November-1 December 1970.* London, Institute of Petroleum, 1971.
82. _______. *Oil and Salt Marsh Soil.* In: Cowell, E.B., editor. *Proceedings of the Symposium on the Ecological Effects of Oil Pollution on Littoral Communities, London, 30 November-1 December 1970.* London, Institute of Petroleum, 1971.
83. _______. *The Effects of Oil on Plants. Environmental Pollution*, 1, pp. 27-44, 1970.

84. _______. *Growth Stimulation Following Oil Pollution.* In: Cowell, E.B., editor. *Proceedings of the Symposium on the Ecological Effects of Oil Pollution on Littoral Communities, London, 30 November-1 December, 1970.* London, Institute of Petroleum, 1971.
85. _______. *Refinery Effluent.* In: Cowell, E.B., editor. *Proceedings of the Symposium on the Ecological Effects of Oil Pollution on Littoral Communities, London, 30 November-1 December, 1970.* London, Institute of Petroleum, 1971.
86. Rutzler, K., and Sterrer, W. *Oil Pollution: Damage Observed in Tropical Communities along the Atlantic Seaboard of Panama. BioScience*, 20, pp. 222-224, 1970.
87. Clark, R.B. *Reports from Rapporteurs.* In: Hepple, P., editor. *Water Pollution by Oil: Proceedings of a Seminar Held at Aviemore, Inverness-shire, Scotland, 4-8 May 1970,* pp. 366-370. London, Institute of Petroleum, 1971.
88. Dunbar, M.J. *Ecological Development in Polar Regions. A Study in Evolution.* Prentice-Hall, Englewood Cliffs, N.J., 119 p., 1968.
89. Connell, D.W. *The Great Barrier Reef Conservation Issue–A Case History. Biological Conservational*, 3, pp. 249-254, 1971.
90. Johannes, R.E., Maragos, J., and Coles, S.L. Oil damages corals exposed to air. Marine Pollution Bulletin, 3, 29-30, 1972.
91. Reish, D.J. The effect of oil refinery wastes on benthic marine animals in Los Angeles Harbor, California. Symposium Commission internationale exploration scientifique Mer Mediterranee, Monaco, *1964*, 355-361, 1965.
92. Baker, J.M. The effects of a single oil spillage. *In:* Cowell, E.B., editor. Proceedings of the symposium on the ecological effects of oil pollution on littoral communities, London, 30 November–1 December, 1970. London, Institute of Petroleum, 1971.
93. Milgram, J.H. Technological aspects of the prevention, control and cleanup of oil spills in the ocean. Report to the Energy Policy Project, 1973.
94. Crapp, G.B. Laboratory experiments with emulsifiers. *In:* Cowell, E.B., editor. Proceedings of the symposium on the ecological effects of oil pollution on littoral communities, London, 30 November-1 December, 1970. London, Institute of Petroleum, 1971.
95. George, J.D. The effects of pollution by oil and oil-dispersants on the common intertidal polychaetes *Cirriformia tentaculata* and *Cirratulus cirratus.* Journal of Applied Ecology, 8, 411-420, 1971.
96. Canevari, G.P. Oil spill dispersants–current status and future outlook. *In:* American Petroleum Institute. Proceedings of joint conference on prevention and control of oil spills, Washington, D.C., June 15-17, 1971. pp. 263-270. Washington, D.C., American Petroleum Institute, 1971.
97. Gumtz, G.D., and Meloy, T.P. Froth flotation cleanup of oil-contaminated beaches. *In:* American Petroleum Institute. Proceedings of a joint conference on pollution and control of oil spills, Washington, D.C. June 15-17, 1971. pp. 523-531. Washington, American Petroleum Institute, 1971.
98. Straughan, D. Breeding and larval settlement of certain intertidal invertebrates in the Santa Barbara Channel following pollution by oil. *In:* Allan Hancock Foundation. Biological and oceanographical survey of the Santa

Barbara Channel oil spill 1969-1970. Vol. 1, compiled by D. Straughan. pp. 223-244. Allan Hancock Foundation, University of Southern California, 1971.

99. Chan, G.L. The effects of the San Francisco oil spill on marine organisms. Part I. College of Marin, Kentfield, California, 78 pp, 1972.

100. Blumer, M., Sass, J., Souza, G., Sanders, H., Grassle, F., and Hampson, G. The West Falmouth oil spill. Technical Report Woods Hole Oceanographic Institution, No. *70-44*, 53 p., 1970.

101. Copeland, B.J., and Steed, D.L. Petrochemical waste systems. *In:* Odum, H.T., Copeland, B.J., and McMahan, E.A., editors. Coastal ecological systems of the United States, Rept. to the Federal Water Pollution Control Administration (RFP 68-128), 1969.

102. Baker, J.M. Biological effects of refinery effluents. *In:* American Petroleum Institute. Proceedings of a joint conference on prevention and control of oil spills, Washington, D.C., March 13-15, 1973. pp. 715-724. Washington, American Petroleum Institute, 1973.

103. Mackin, J.G. A study of the effect of oilfield brine effluents on biotic communities in Texas estuaries. Texas A. and M. Research Foundation Project 735, 1971.

104. Hohn, M. The use of diatom populations as a measure of water quality in selected areas of Galveston and Chocolate Bay, Texas. Publications of the Institute of Marine Science University of Texas, *6*, 206-212, 1959.

105. Odum, H.T., Cuzon du Rest, R.P., Beyers, R.J., and Allbaugh, C. Diurnal metabolism, total phosphorus, Ohle anomaly, and zooplankton diversity of abnormal marine ecosystems of Texas. Publications of the Institute of Marine Science University of Texas, *9*, 404-453, 1963.

106. Chambers, G.V., and A.K. Sparks. An ecological survey of the Houston Ship Channel and adjacent bays. Publications of the Institute of Marine Science University of Texas, *6*, 213-250, 1959.

107. Farrington, J.W., and Quinn, J.G. Petroleum hydrocarbons in Narragansett Bay. I. Survey of hydrocarbons in sediments and clams (*Mercenaria mercenaria*). Estuarine and Coastal Marine Science, *1*, 71-79, 1973.

108. Jeffries, H.P. A stress syndrome in the hard clam, *Mercenaria mercenaria.* Journal of Invertebrate Pathology, *20*, 242-251.

109. Storrs, P.N. Petroleum inputs to the marine environment from land sources. Engineering-Science, Inc., unpublished manuscript, 1973.

110. International Decade of Ocean Exploration. Baseline studies of pollutants in the marine environment and research recommendations. The IDOE Baseline Conference, May 24-26, 1972, 54 p., 1972.

111. Blumer, M. Submarine seeps: Are they a major source of open ocean oil pollution. Science, *176*, 1257-1258.

112. Blumer, M. Personal communication, 1973.

113. Atema, J., and Stein L. Sublethal effects of crude oil on the behavior of the American Lobster. Technical Report Woods Hole Oceanographic Institution, No. *72-74*, 1972.

114. Jacobson, S.M., and Boylan, D.B. Effect of seawater soluble fraction of kerosene on chemotaxis in a marine snail, *Nassarius obsoletus*. Nature, 241, 213-215.

115. Suess, M.J. Polynuclear aromatic hydrocarbon pollution of the marine environment. 5 p. (Preprint of a paper presented at the F.A.O. Technical Conference on marine pollution and its effects on living resources and fishing, Rome 9-18 December 1970). (FIR: MP/70/E-42)
116. National Academy of Sciences. Particulate polycyclic organic matter. National Academy of Sciences, Washington, D.C., 1972.
117. Barry, M.M., Yevich, P.P., and Thayer, N.H. Atypical hyperplasia in the soft-shell clam *Mya arenaria.* Journal of Invertebrate Pathology, 17, 17-27, 1971.
118. Barry, M.M., and Yevich, P.P. Incidence of gonadal cancer in the quahaug *Mercenaria mercenaria.* Oncology, 26, 87-96, 1972.
119. Murphy, T.A. Environmental effects of oil pollution. Journal of Sanitary Engineering, Division Proceedings of the American Society of Civil Engineers, *8221*, 361-371, 1971.
120. Foster, M., Charters, A.C., and Neushul, M. The Santa Barbara oil spill: Part 1 - Initial quantities and distribution of pollutant crude oil. Environmental Pollution, 2, 97-114, 1971.
121. McAuliffe, C.D. The environmental impact of an offshore oil spill. Chevron Oil Field Research Company, La Habra, Calif., 1973.
122. Mackin, J.G. A review of significant papers on effects of oil spills and oil field brine discharges on marine biotic communities. Texas A. and M. Research Foundation Project 737, 1973.
123. Wilson, K.W., Cowell, E.B., and Beynon, L.R. The toxicity testing of oils and dispersants: a European view. *In:* American Petroleum Institute. Proceedings of a joint conference on prevention and control of oil spills, Washington, D.C., March 13-15, 1973. pp 255-261. Washington, American Petroleum Institute, 1973.

Part Two

Technological Aspects of the Prevention, Control, and Cleanup of Oil Spills

Chapter One

Introduction

A vast amount of material has been written on oil pollution and its effects during the past five years. Uncertainty is a general feature of most of the reports. For example, published estimates of the total annual influx of oil into the oceans of the earth vary from 1.64 million tons to 10 million tons (1).[a]

If 10 million tons of oil were put into the sea annually and uniformly distributed over the world's oceans, the influx rate would be only six ounces of oil per square mile per day. Most oil pollution problems are due not to the overall spillage, but to the spillage of a large amount of oil in a small area in a short time.

Porricelli *et al.* (1) have estimated the quantities of oil reaching the ocean from various sources. Their estimate, as shown in Figure 1-1 and Table 1-1, is subjective at best. Automobile crankcase oil disposal is given as the largest source of oil pollution in the oceans, and Porricelli's method of calculating this pollution (1) is an example of this subjectivity. Porricelli estimates that on January 1, 1970 there were 207 million automobiles, trucks and buses in the world. Then he assumes that each vehicle changes oil twice a year and that each oil change involves an average of 5.5 quarts of oil, thus generating 569 million gallons of used crankcase lubricating oil each year. He then estimates (without substantiating evidence) that 75 percent of this oil-425 million gallons–enters the sea annually. It is probably impossible to estimate accurately the percentage of used crankcase oil that enters the sea. One difficulty with questionable estimates is that they are often used authoritatively in subsequent reports. For example, the environmental impact statement of the Maritime Administration Tanker Construction Program (2) uses Porricelli's estimate of the annual ocean pollution by motor vehicle lubricating oil.

The extent of ocean pollution by natural seepage of oil is also uncertain. An accurate estimate of the extent of natural seepage would be

---

[a]Numbers in parentheses refer to bibliography listed after the text of this report.

Table 1-1. Estimated Annual Pollution of the Oceans by Man*

Marine Operations	Metric Tons	Percent
Tankers		
LOT tank cleaning operations	265,000	5.41
Non-LOT tank cleaning operations	702,000	14.34
Discharge due to bilge pumping, leaks and bunkering spills	100,000	2.04
Vessel casualties	250,000	5.11
Terminal operations	70,000	1.42
Tank Barges		
Discharge due to leaks	20,000	0.41
Barge casualties	32,000	0.65
Terminal operations	18,000	0.38
All Other Vessels		
Discharge due to bilge pumping, leaks and bunkering spills	600,000	12.25
Vessel casualties	250,000	5.11
Offshore Operations	100,000	2.04
Nonmarine Operations		
Refineries and petrochemical plants	300,000	6.12
Industrial machinery	750,000	15.31
Highway motor vehicles	1,440,000	29.41
Total	4,897,000	100.00

*Source: Porricelli, J.D., Keith, V.F. and Storch, R.L., "Tankers and the Ecology," Transactions of the Society of Naval Architects and Marine Engineers, Vol. 79, 1971.

valuable, since a comparison of the natural seepage rate with the manmade spill rate would help in assessing the potential damage. Divers in the Santa Barbara Channel have reported numerous natural seeps in that area. The oil from some of the seeps is heavier than water and does not float. As might be expected, the oil influx into the sea from non-floating natural seepage is unknown.

Three sources of oil pollution will be considered in this report—tankers, wells, and pipelines. Refineries are another source of oil pollution, and many references list them as a major source (1). However, most of the reported refinery pollution came from a single spill (4). If this one spill is excluded from the data, refineries are found to be only a minor source of oil pollution (4).

Petroleum typically undergoes ten to fifteen transfers from its production to its consumption (3), so a given quantity of oil has many opportunities to be spilled. The Georges Bank petroleum study (4) concludes that 75 percent of all spills are caused by human error and 25 percent by

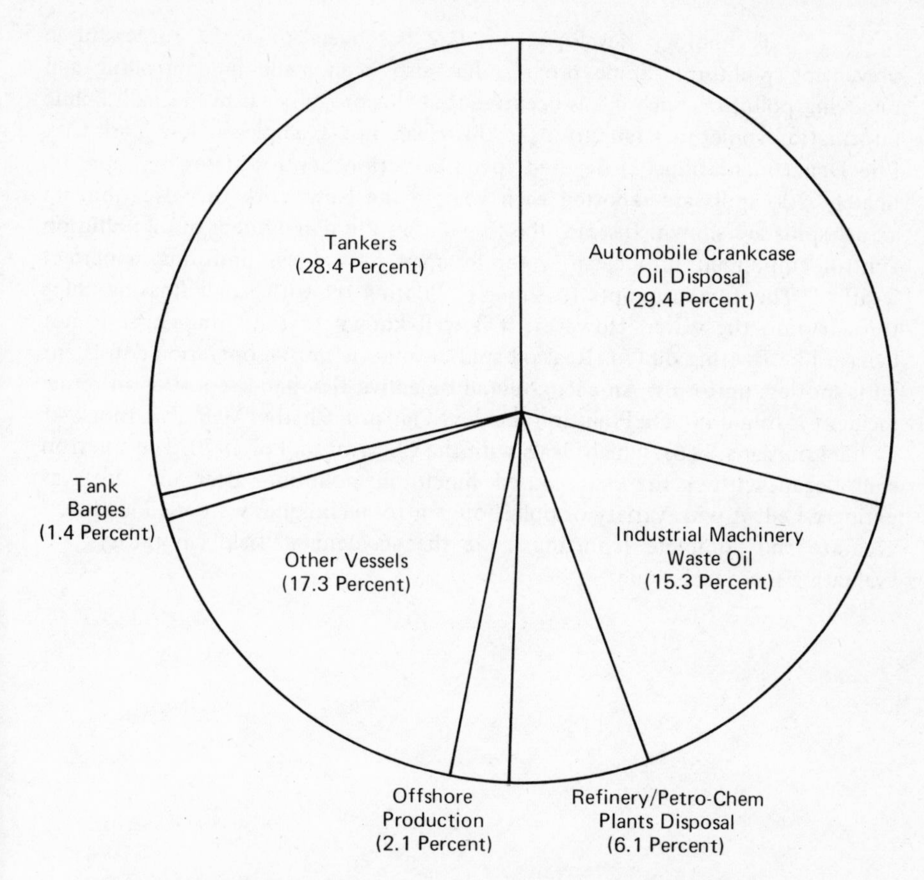

Source: Porricelli, J.D., Keith, V.F. and Storch, R.L., "Tankers and the Ecology," Transactions of the Society of Naval Architects and Marine Engineers, Vol. 79, 1971.

Figure 1-1. Sources of Oil Pollution in the Oceans.

technical failure. However, 75 percent of the total volume of spilled oil comes from technical failure and only 25 percent from human error. This indicates that spills due to technical failure are, in general, much larger than spills due to human error. It is important to realize, then, that although most of the spillage volume could be eliminated if near-perfect technology could be developed, the actual number of spills would be reduced by only 25 percent. In many instances, the damage caused by an oil spill would be reduced only slightly if the spilled volume were less. It is nearly as hard to clean up one quart of oil from a stretch of beach as it is to clean up a gallon from the same spot. Thus, if the spills due to technical failure were eliminated, eliminating 75 percent of the spilled oil, it is likely that much more than 25 percent of the damage would remain.

Technology developed to date has been somewhat successful in preventing pollution. Some progress has also been made in controlling and removing pollution once it has occurred, but this progress has been small. Public information sometimes fails to make this clear. For example, a New York City Fire Department film (3) devoted to oil pollution control states that approximately 200 spills are reported each year in the New York City area, but no actual spills are shown. Instead, the film shows the deployment of oil pollution control equipment into clean water by men in spotless uniforms in perfect weather. The film attempts to simulate floating oil with small floating chips thrown onto the water. However, it is well known that floating chips do not behave like floating oil (5). Real oil spill cleanup is not an operation consistent with spotless uniforms. An accurate and objective first-hand report of an actual incident is found in "Oil Pollution Incident Platform Charlie, Main Pass Block 41 Field, Louisiana," (6), which deals with the Chevron spill of 1970. The Chevron spill began with a fire and caused much oil pollution after the fire was extinguished. A wide variety of pollution control techniques were used, and thus accurate and complete reporting made the accident a "field laboratory" for evaluating these techniques.

Chapter Two

Spill Prevention

SPILL PREVENTION FROM TANKERS

The statement in the Georges Bank Study (4) that three-quarters of the oil spilled in accidents comes from the 25 percent of accidents due to technical failure does not include the pollution caused by pumping ballast water overboard. This is because the study was concerned only with accidents, whereas the pumping of ballast water is "intentional." The deballasting of tankers accounts for approximately 70 percent of the tanker-caused oil pollution of the oceans today (1). Therefore, with the possible exception of pollution from highway motor vehicles, ballast water discharge is the largest source of unnatural influx of oil into the sea.

Oil tankers usually carry products in only one direction and return voyages are made without cargo. For a ship with the size and weight distribution of an oil tanker, going to sea unloaded is neither comfortable nor safe. For this reason, these ships fill some of their tanks with sea water (called ballast water) when making return voyages. The oil remaining in the bottoms of the tanks and adhering to the walls of the tanks mixes with the ballast water on these voyages and, unless special equipment and procedures are used, the oily water is discharged directly into the sea prior to the tanker's arrival at port. The first two items in Table 1-1 relate to the discharge of oily ballast, the first relating to tankers using a pollution-minimizing procedure called "load-on-top," and the second relating to tankers not using this procedure. The deballasting of tankers usually takes place about twenty or thirty miles offshore, and much of the associated oil never reaches shore, so the coastline is far less polluted by ballast water discharge than is the ocean itself.

New standards for high seas deballasting were recently established by the Inter-Governmental Maritime Consultative Organization (IMCO). If ratified by the necessary fifteen nations, the standards will require all ships over 150 tons to use load-on-top procedures and to limit all discharges to waters over fifty miles from shore. Tankers built after 1975 will be required to have segregated ballast tanks to avoid the mixing of oil and water.

Load-on-Top Procedure

Many of the newer tankers traveling long routes employ the so-called load-on-top (LOT) procedure for minimizing pollution from oily ballast water discharge. To accomplish the load-on-top procedure, the following steps (shown schematically in Figure 2-1) have been widely adopted.

1. After discharging the cargo and proceeding out to sea, several of the ship's tanks are filled with sea water to maintain stability.
2. The remaining empty cargo tanks are cleaned of the oil that has remained on the walls and bottom of the tank by directing jets of sea water from high pressure revolving nozzles over the entire tank surface.
3. The oily water from the tanks that have been washed down inside is pumped into an empty cargo tank or a specially designated "slop tank."
4. At this point, having been washed relatively clean, the empty cargo tanks are slowly filled with sea water.
5. By this time, the tanks initially filled with sea water have been allowed to stand for a considerable time, and the oil residue that was initially mixed with the ballast water has risen to the top of the tank. The relatively clean water beneath the oil in these tanks is slowly discharged overboard until the oil level is near the bottom of the tanks. As much of this oil as possible is pumped into the slop tank, and the cargo tanks are then washed down with the high-pressure water system.
6. The oily water mixture that accumulated in these tanks during washing is then pumped into the slop tank.
7. The mixture in the slop tank is now allocated to settle, with the oil rising to the top. The water is discharged slowly until the oil-water interface approaches the bottom of the tank, whereupon all pumping operations are stopped.
8. Before arrival at the loading dock, the clean ballast water is discharged, leaving all the oil in the slop tank. New oil is taken aboard and is loaded on top of the oil remaining in the slop tank.

The load-on-top procedure is not suitable for all tanker operations. Two to five days are required for gravity separation of the oil and water, and the motions of the ship must be quite mild to allow separation to take place. Many tankers travel routes too short or waters too rough to allow effective use of the load-on-top procedure. However, where the load-on-top procedure can be effective, it does much to reduce the discharge of oily ballast into the sea. Approximately 86 percent of the world's tanker capacity uses the load-on-top procedure, according to a representative of one major oil company.

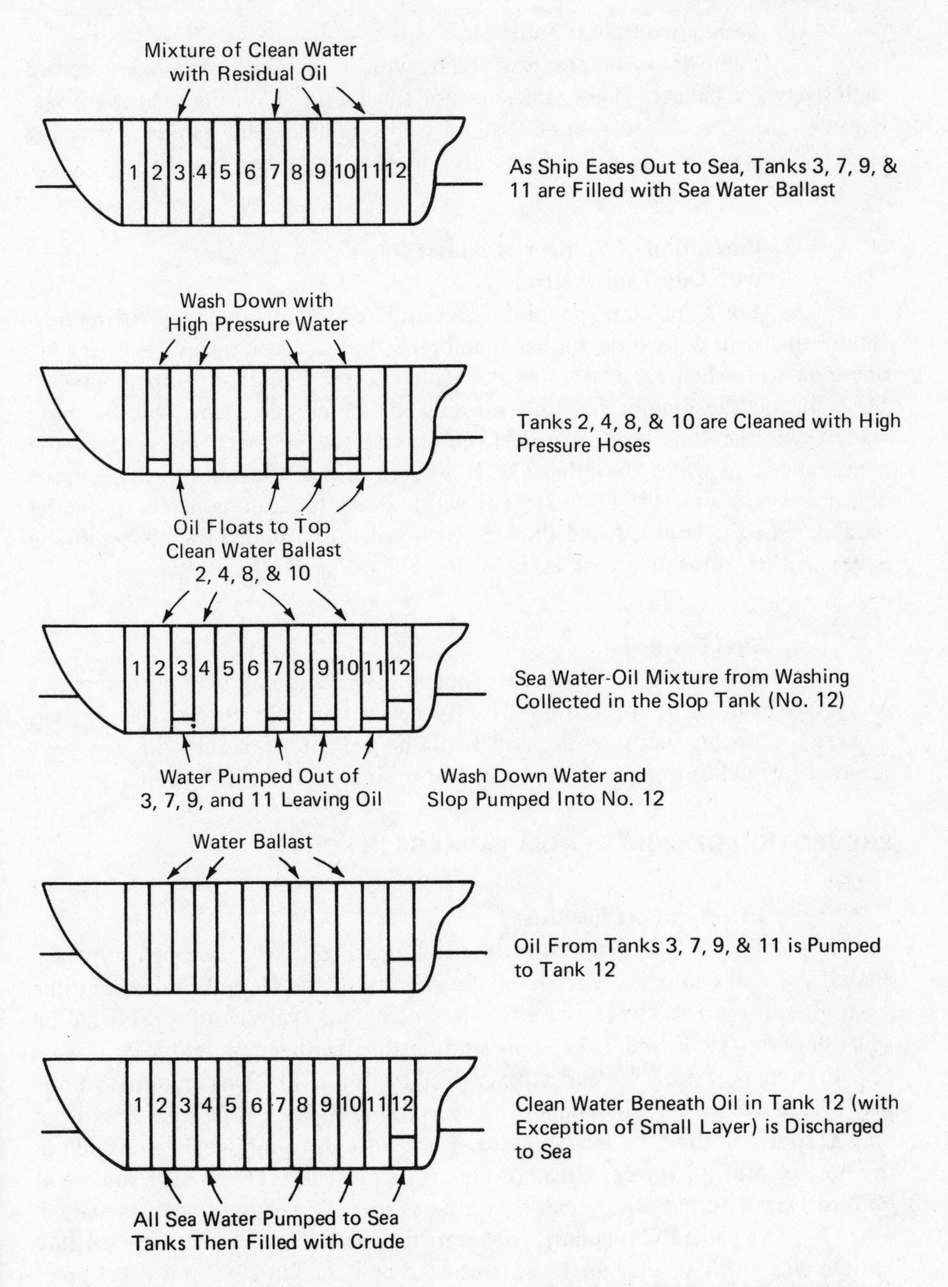

Source: Maritime Administration Tanker Construction Program, Draft Environmental Impact Statement, Volume I.

Figure 2-1. The Load-On-Top Procedure.

Segregated Ballast Tanks

Some tankers are now built with a number of tanks reserved exclusively for ballast. These tanks are not filled with oil during a cargo voyage. Because the use of segregated ballast tanks reduces the tanker's carrying capacity, this practice is avoided wherever possible by tanker operators.

Prevention of Contact of Ballast Water with Oily Tank Walls

For some time, people have considered the possibility of fitting each cargo tank with a flexible rubber membrane that would be pressed up against one tank wall when cargo was carried. Ballast water would be pumped between the wall and the membrane, thus pressing the membrane against another tank wall, when the tank was in ballast. Oil would always touch one side of the membrane, and water the other (1). However, serious research on this concept did not begin until 1972 (13). The work done to date indicates potential feasibility of the concept, and the U.S. Navy recently installed such a membrane in one of their ships on a trial basis.

Slop Burning

Porricelli (1) discusses the possibility of burning slop oil on tankers using the load-on-top procedure. No further research in this area is known, however, probably because slop oil is discharged ashore along with the next cargo of oil, and burning is not practical for ordinary oily ballast water.

PREVENTION OF SPILLS FROM TANKERS IN PORT

Effect of Port Facilities

The Georges Bank Study (4) contains a relatively thorough statistical analysis of spills into the oceans in the vicinity of the United States coastline from all sources for 1971, as well as less complete analyses of worldwide oil spills between 1957 and 1971. This study estimates that 500,000 gallons were spilled from port transfer and storage facilities in 1971. This amounts to one part spilled per million parts handled. It was found that ten times this amount of oil was spilled in the New England area. The report also contains data on spills at the port of Milford Haven, Great Britain, for 1963-1969. The average spillage at Milford Haven during this period, measured as a fraction of the total oil handled, was 1.8 parts per million (ppm). However, for the last two years reported, the spillage was much smaller; in 1968 it was 0.5 ppm, and in 1969 it was 0.4 ppm. Milford Haven is a very large new port. In 1969, with an average daily throughput of 745,000 barrels of oil, the total quantity spilled was 4,050 gallons. The total average spillage at Milford Haven between 1963 and 1969 is unrepresentatively high because of some problems in 1967 when 73,000 gallons were spilled. If 1967 is excluded from the Milford Haven data, the average amount of oil spilled is 0.6 ppm.[a] This figure indicates that in a new port, with modern design and new equipment, spillage can be somewhat less than that in outdated facilities such as those in the New England area.

High Oil Level Alarms

The control of cargo loading on most older tanks is a manual process. For this reason, it is important for cargo loading workers to know when tanks become nearly full to prevent tank overflow and consequent pollution. Alarm systems that make a sound when the oil nears the top of a tank have been under development for some time (14). However, as is often the case with new technical devices, the enthusiasm in published papers is not justified by subsequent events. Leonard (14) reported an extremely successful development project for these alarm systems in 1971. But in a 1973 investigation, it was determined that oil companies found alarms unreliable and were attempting to have better ones developed. Thus, two years after the report of successful development of these alarm systems, users find them unreliable. This is not unusual in marine technology. The marine environment is extremely hard on any instrument, and marine practices do not allow the frequent testing and maintenance of pieces of equipment. This is in sharp contrast to practices in the air transportation industry.

Control of Cargo Handling

On some new tankers, control of loading and unloading operations emanates from on-board monitoring devices which show the amount of oil in all tanks at all times (16). If these systems prove reliable, they will reduce the pollution risk from tankers in port. However, since the usual tanker lifetime is about twenty years, many older tankers without such systems will be making port visits for a long time to come. One way to minimize pollution risks from older tankers is to arrange shoreside cargo handling control on special loading and unloading docks with monitoring devices of their own.

Personnel Practices to Minimize Spills

Most oil companies use special procedures and personnel training to minimize spills on vessels in port instead of developing and relying on complicated technical devices (16 and 17). Most companies carry about half of their oil in chartered vessels. In some cases, retired ship captains are put aboard vessels when they come into port to serve as pilots and help prevent a spill. The pilot checks all operations and offers advice for correcting deficiencies that could lead to a spill. This program has been very successful and has been well received by most regular ship captains.

[a]Editor's note: This figure does not reflect the August 1973 spill at Milford Haven, when 2,500 to 3,500 tons were spilled. That spill is described in the Introduction to this volume.

SPILL PREVENTION FROM TANKER ACCIDENTS ON THE SEAS

Figure 2-2 shows the amount of crude oil transported by sea annually for the past few years, with projections for future years. The past data agrees with other estimates of oil transported (4). At the present time, about 1.5 billion metric tons of oil are transported by sea each year. As shown in Table 1-1, approximately 250,000 tons of oil per year are spilled due to tanker casualties. It is likely that this is an underestimate, because some spills due to tanker casualties far from land are not reported. Even so, pollution due to tanker casualties is frequently in the public eye because of the large amounts of oil that are sometimes spilled. Porricelli (1) provides a comprehensive analysis of tanker casualties for 1969 and 1970 covering 1,416 of them, including 269 that caused pollution. Figure 2-3 shows a breakdown of the polluting incidents in terms of the type of casualties. Two hundred nine of the 269 polluting incidents occurred in harbors, harbor entrances, coastal zones, and at piers. Except for a few casualties at unknown locations, most of the remaining sixty casualties occurred

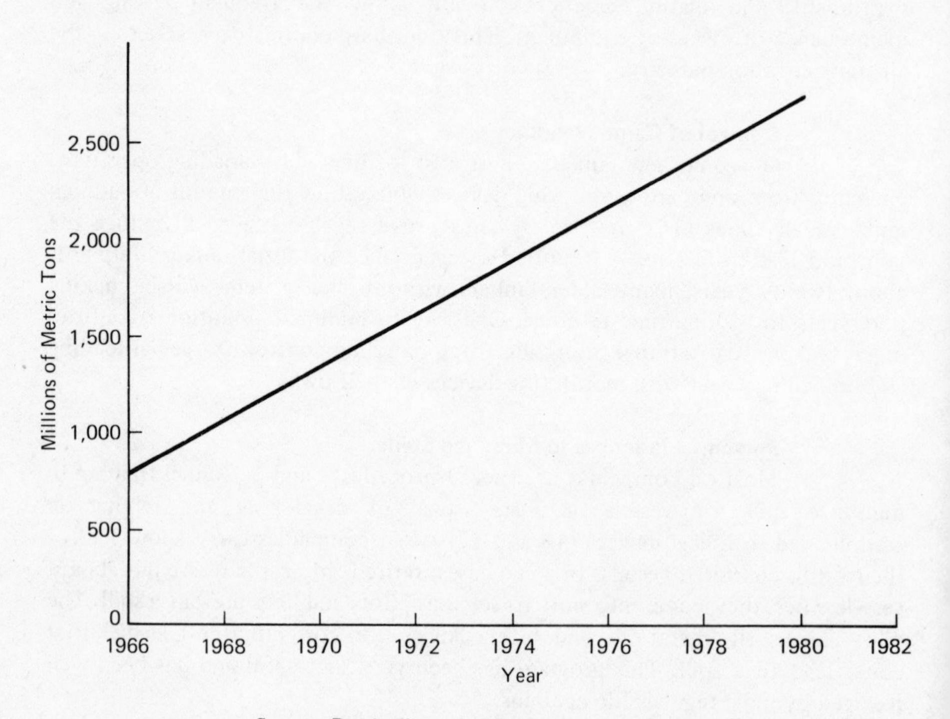

Source: Porricelli *et al.*, "Tankers and the Ecology."

Figure 2-2. The Amount of Oil Transported at Sea Annually.

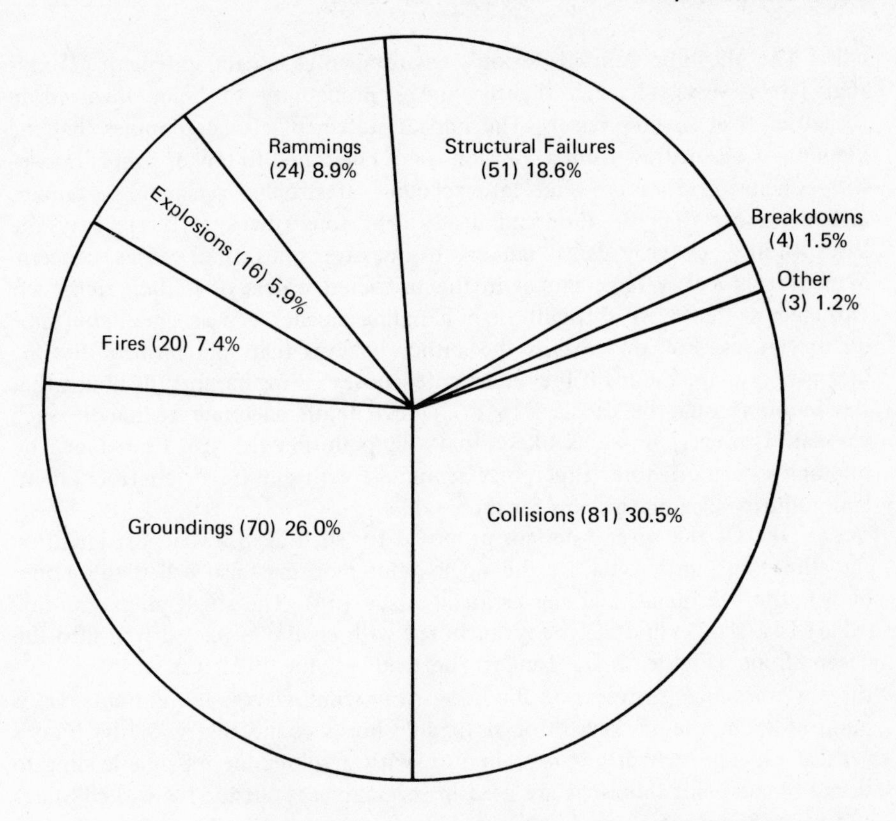

Source: Porricelli *et al.*, "Tankers and the Ecology."

Figure 2-3. Type, Number, and Percentage of the 269 Polluting Tanker Casualties.

on the high seas, and of these forty-one were due to structural failure. Hence, it can be concluded that nearly all tanker casualties occur near land, and most of the remaining casualties are due to structural failures. The major sources of pollution-causing incidents are groundings, collisions, and structural failures. Groundings, of course, can only occur in the shallow water near land. Collisions and rammings are most likely to occur where there is heavy traffic–in harbors, harbor entrances, and approaches.

Two oil company representatives, interviewed (15 and 16) to determine the industry's view on the relative oil spill probabilities of large and small tankers, believe that larger tankers will cause fewer accidents because fewer ships will be required. This judgment is speculative, because hard data do not exist. The Maritime Administration's environmental impact statement (2) says that larger vessels have a slightly higher probability of being involved in casualties than smaller vessels. The impact statement also determines that the amount of oil outflow from a collision is not correlated to tanker size. However, this conclusion does not take into account catastrophic accidents to tankers carrying more than 80 thousand deadweight tons (dwt). It is precisely the susceptibility of very large tankers to massive spills that causes concern. Maneuvering a very large tanker in the restricted waters of harbors and their entrances is far more difficult than handling smaller vessels under the same circumstances. For this reason the author believes that spill probabilities are increased by larger ships if they must enter most existing harbors (assuming that the harbors could be dredged to a channel depth adequate to handle such vessels). However, it seems likely that the pollution risk can be reduced by building special offshore "superports" complete with ship traffic control systems and pollution cleanup systems for these tankers.

Of the several designs proposed for such deep water port facilities, the three most promising are the single-point mooring (also called the mono-buoy), the sea island, and the artificial island (54). The single-point mooring (SPM) is a flat, cylindrical buoy anchored with chains to piles driven into the ocean floor (Figure 2-4). Tankers berthed at the SPM can rotate 360°, thereby minimizing stresses on the buoy from wind, waves, and currents. Oil is pumped from the tankers through flexible hoses connecting with the buoy's vertical pipeline, which connects in turn with a submarine pipeline leading to shore. Single-point moorings are used in several places outside the United States and have shown surprising durability in high seas and winds. Tankers may not be able to berth at SPM's under such conditions, however, because they often require assistance from tugboats incapable of maneuvering a tanker in very rough weather. Plans to install SPM's in the Gulf of Mexico center around sites fifteen to thirty miles offshore.

A sea island (Figure 2-5) is a platform mounted on pilings with berthing facilities on both sides and connected to shore by a submarine pipeline. Sea islands are efficient, permitting a high rate of oil transfer. Because they keep tankers in a fixed position, they must be constructed parallel to the prevailing winds. They require tug assistance for berthing and cannot usually be used in waves greater than three feet. The deep water port at Milford Haven, Wales includes a sea island.

An artificial island (Figure 2-6) would be exactly what its name implies. By building an enclosed rock-fill dike and filling the enclosure with concrete, a large permanent island would be formed. A breakwater would surround the island to protect it from high seas. Berthing facilities could be installed on all sides, affording shelter from high waves and winds on any side. The island could include storage facilities–for dry bulk commodities as well as for oil–and docking facilities for small shuttle vessels to take cargo to shore. Barges and pipelines might also provide transportation to shore. There are no

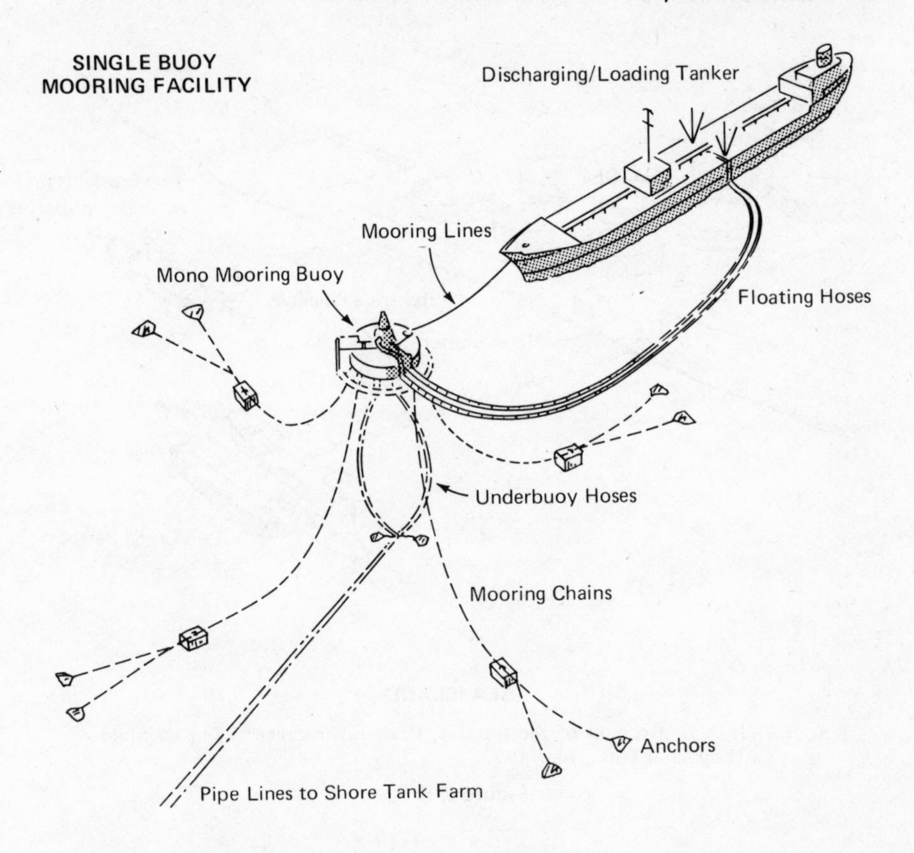

Source: U.S. Department of the Interior, Draft Environmental Impact Statement on Deepwater Ports, June 1973.

Figure 2-4.

artificial islands of this type at present, and their future development is speculative because of the extremely high cost.

All of these facilities offer the potential of reducing oil spill risks from very large tankers because they eliminate the need for difficult maneuvers in heavy traffic. They are also likely to be used to moor supertankers in United States ports, because existing harbors are too shallow and dredging operations to deepen them are generally considered too difficult as well as harmful to the environment.

Tanker Structures and Structural Standards

IMCO has approved standards limiting the maximum single tank size on tankers to 40 thousand cubic meters (10.6 million gallons). However, in

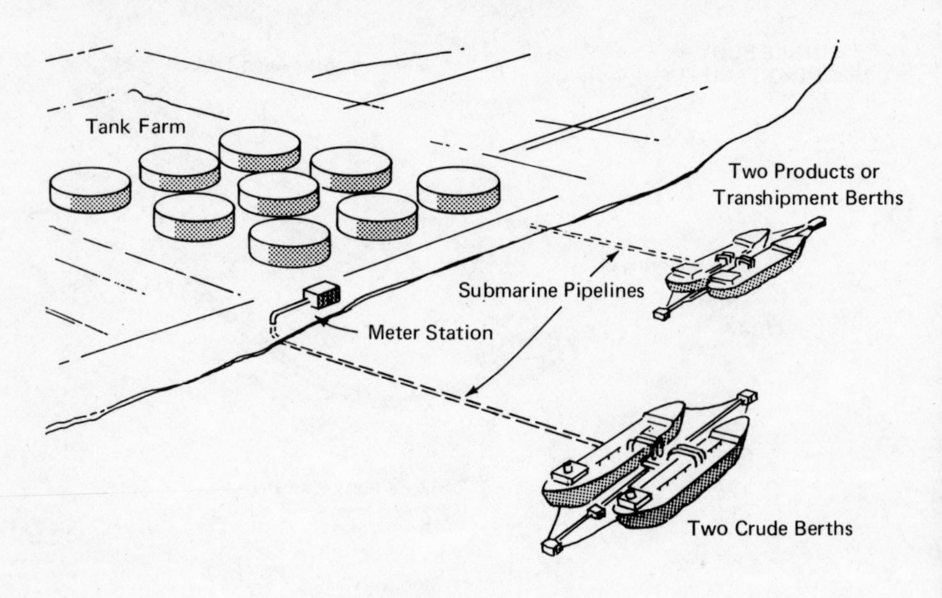

SEA ISLAND

Source: U.S. Department of the Interior, Draft Environmental Impact Statement on Deepwater Ports, June 1973.

Figure 2-5.

many casualties, all the oil from at least one tank escapes into the sea. A large tanker casualty can result in massive pollution even if the tanker is built to the IMCO standards and only a single tank is ruptured. The pollution risk would be diminished if the allowable maximum tank size were reduced. However, there is oil industry opposition to a reduction in tank size; in fact, the industry wants larger tanks because they offer economies of scale. Thus, improvement in tanker structures and the development of adequate structural standards is especially desirable. The large number of structural failures of tankers indicates that existing structural standards are inadequate.

Present-day ocean engineering knowledge is adequate for developing tanker structural standards to reduce the risk of structural failure at sea in any storm. Since ocean waves are random, structural survival standards for severe conditions must be developed on the bases of probability theory. It is not possible to develop a fail-safe structure in a sea of unlimited severity. Rather one must design tankers according to the failure probability for each trade route at each time of year, since storm probabilities depend on time of year.

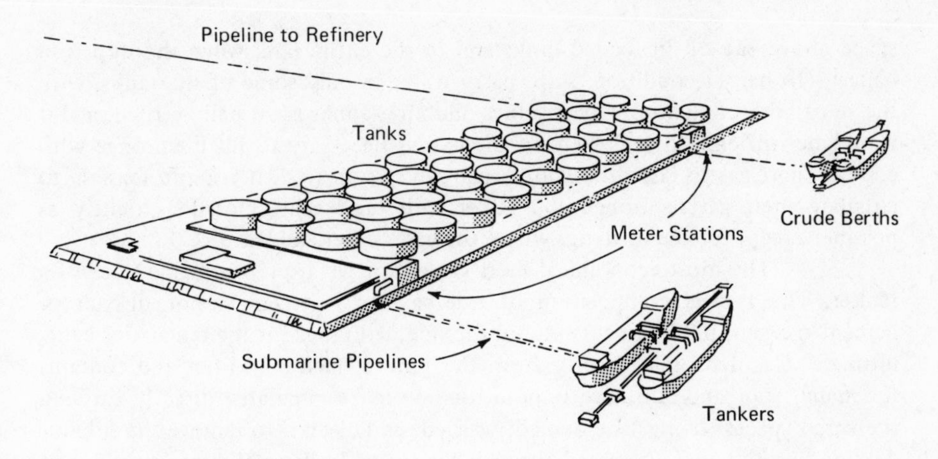

ARTIFICIAL ISLAND

Source: U.S. Department of the Interior, Draft Environmental Impact Statement on Deepwater Ports, June 1973.

Figure 2-6.

The enormous size of some modern tankers means that uneven tank filling during loading or unloading could result in great enough stresses to break the ship even in completely calm water. The oil industry is undertaking studies (16) to determine load stresses as a function of weight distribution to establish loading and unloading sequences that would minimize the probability of ship breakage during the on- and off-loading of cargo.

Tank Inerting Systems for Minimizing Explosion Risks

Porricelli (1) studied explosions of thirty-two tankers, sixteen of which caused pollution. When hydrocarbons in a tank evaporate into the air, there is an explosion risk if the oxygen percentage in the air exceeds 11 percent. Ordinary air contains 21 percent oxygen. Twenty-one of the explosions Porricelli looked at originated in the cargo spaces. Fifteen occurred when the ships were in the ballast condition, in relatively empty tanks where a large amount of explosive air-oil vapor could exist. Seven of the tankers that exploded were over 100,000 dwt, and three of them now rest on the ocean bottom.

The object of an inerting system is to reduce the oxygen level to well below 11 percent by filling all cargo tanks with inert gas. This applies to the space above the oil in loaded tanks and to the entire tank when the ship is in ballast. (In ballast condition, ships carry water in only some of the tanks, with the remainder containing oily residues and air). Tanks are usually vented, and it would be difficult to seal them to the extent necessary to fill them once with enough inert gas to last throughout each voyage. Instead, it is more feasible to circulate inert gases through the tanks with tank vents closed as tightly as possible except for the openings where the gases enter and leave.

The most economical inert gas is flue gas from the engine stack of a tanker. The typical composition of flue gas is 13 percent carbon dioxide, 4 percent oxygen, 0.25 percent sulphur dioxide, with most of the remainder being nitrogen. Gas delivered directly from the tanker stack is too hot and contains too much soot and corrosive sulphur dioxide to be circulated directly through the cargo spaces. It must be cooled, washed, and cleaned to remove the sulphur dioxide before being circulated through the tanks. In Porricelli's analysis (1), no tanker carrying an inerting system exploded.

Improved Electronic Navigation and Operating Practices

Nearly all tanker groundings occur because the vessel is "lost" just prior to the grounding. There are many examples of this.

When the *Torrey Canyon* ran up on Seven Stones Reef, U.S. Hydrographic Office Chart #H.O. 450 was on the chart table on the bridge of the vessel (19). The *Torrey Canyon* hit Pollard Rock on the reef because the people on the bridge did not know precisely where the ship was in relation to the chart until it was too late.

Cerberus Rock is the only rock far from shore in all of Chedabucto Bay, Nova Scotia. The *Arrow* hit this rock in February 1970, the result being massive pollution of the bay by her cargo of Bunker C oil. Personnel on the bridge of the vessel apparently had no idea of their exact location, since the rock was easily avoidable.

These accidents and most other groundings could have been avoided by good piloting. For the region of the *Torrey Canyon* accident the principal piloting document is "The Channel Pilot," which specifically warns that the route taken by the *Torrey Canyon* is dangerous for large vessels. There was no copy of "The Channel Pilot" aboard the ship, even though this voyage, like her previous voyage, took her into the English Channel. In the case of the *Arrow*, Cerberus Rock is marked by a large buoy with a flashing light. Such a buoy could easily be seen by eye in good weather and by radar in nearly all conditions.

It is quite easy for the operators of a large ship to become bored and complacent during long voyages. For this reason, good operating practices cannot be stressed enough.

Satellite navigation offers the possibility for improved electronic navigation. The degree of improvement is a matter of opinion, and experimental systems have not lived up to some expectations. To make such a system work, satellites sending out radio waves must always be within range of the vessel. Electronic devices can determine the ship location from the radio signals. At the present time, there are only a limited number of these satellites, so that position-fixing by satellite is possible for each vessel only at certain times of each day. More of these satellites would provide a worldwide network allowing continuous, accurate, essentially automatic position-fixing.

Collision Avoidance Radar

The maneuverability of a vessel generally decreases as the vessel size and weight are increased. Because supertankers are now the largest vessels on the seas, by their very size they must be relatively unmaneuverable. Moreover, they are underpowered. To understand this, one can visualize a 15-foot motorboat weighing 1,500 pounds operating in calm conditions. If the same relationship between power and size were applied to this vessel as existed on a 100,000 dwt, one thousand-foot long supertanker having a 20,000 horsepower engine, the motorboat engine would have only 1/60 of a horsepower. Complicated maneuvers would be very difficult with so little power. Much of the economic efficiency of supertankers comes from the fact that they can transport such large amounts of oil with such small power plants. Because of their large size and low power, maneuvers with supertankers take a long time to complete. For example, the stopping distance for a tanker of 200,000 dwt moving at a forward speed of sixteen knots with full stern power applied to stop the ship is about three and a half miles (55).

Recent research has aimed toward the development of collision avoidance radar systems. Such a system combines a radar set and a computer to determine the location of nearby ships and obstructions and to predict the possibility of collisions in advance. The tanker operator can decide on a possible collision avoidance maneuver and indicate this to the computer, which will then compute the trajectory of all vessels, including his own, and show whether the proposed maneuver will indeed avoid collision. If a collision still seems possible after the proposed maneuver, the process is repeated until a safe course is decided upon. Collision avoidance radar presently exists on eleven ships; the major supplier has orders for forty-two more units, of which thirty-nine are for oil tankers.

Harbor Traffic Control

The problem of controlling a large number of ships moving inside a harbor can be ameliorated by a single central traffic control system at the harbor. Such an arrangement would be similar to today's air traffic control for airplanes, but there are a number of particular problems with central harbor traffic control that make it more difficult than controlling aircraft. For example, at nearly all airports, each airplane has a fairly large space where it can maneuver in a holding pattern. This is not the case with large ships in a relatively small harbor, because of restricted space and lack of maneuverability. There is also a sociological problem, since much of maritime law is based on tradition. One aspect of this tradition is that a ship captain is the sole master of his vessel, and that he alone makes all decisions.

In spite of this, some harbor traffic control systems have been put into operation. The most advanced one in the United States is in San Francisco Harbor, where traffic controllers have a radar system that scans the entire harbor. The controllers plot the traffic and advise mariners of potential difficulties. The Coast Guard has been given the authority to make control mandatory there for all vessels, but because of the potential difficulties in carrying this out and the good cooperation the Coast Guard has enjoyed with all ships entering and leaving San Francisco Harbor, control has been left on a voluntary basis.

A somewhat simpler system is used in Seattle, where position reports are supplied to the Coast Guard by the vessels and the Coast Guard plots the positions of the various reporting vessels in the harbor without the use of radar.

A relatively simple system was installed in Chedabucto Bay, Nova Scotia, following the grounding of the ARROW. There, traffic controllers have both radar and voice communications with nearby ships.

Bridge-to-Bridge Communications

A 1971 law requires major vessels to have bridge-to-bridge communications in large harbors. It is well known that collisions can be avoided more easily if vessels have continuous voice communications with each other. A specific radio channel is set aside in each harbor area for voice communications, and all large vessels must be able to receive and transmit on this channel as well as to receive emergency information on a special emergency channel. This requirement applies to all vessels larger than 300 gross tons, passenger vessels larger than 100 gross tons, and all tugboats longer than 45 feet.

SPILL PREVENTION ON OFFSHORE WELLS

Introduction and Role of the U.S. Department of the Interior

Where offshore oil production takes place, oil spills are a chronic problem. This is well summarized in the Chevron spill study (6), which states:

> Spills varying in size from a few gallons to many barrels are endemic to the Gulf of Mexico . . . Oil appears on the Gulf waters from . . . numerous operations connected with the drilling [and] operations of the wells. U.S. Coast Guard's reconnaissance flights report three

to seven pollution incidents every week, and many of these identify with the company or individual responsible. However, the cause of each incident is not generally documented and action to prevent a reoccurrence is not often taken.

Federal control over oil industry offshore practices is exercised by the Department of the Interior, mainly through the Department's United States Geological Survey (USGS). Offshore oil production is actually governed by two levels of federal regulations (46). The higher regulations are those of the Secretary of the Interior and include general leasing and operating regulations. The secondary level of regulations are "orders" issued by the oil and gas supervisors within USGS. The orders for drilling on the Outer Continental Shelf are called OCS orders and are quite specific about technical arrangements. Some provide regional regulations for drilling in the Gulf of Mexico and the Pacific. Others refer specifically to actions of an individual lessee, or to a field.

Both the Secretary's regulations and the operating orders have been changed somewhat since the Santa Barbara oil pollution incident. One change made in the leasing regulations of the Secretary of the Interior provides that full consideration of all environmental factors must be given prior to leasing, including factors concerning aquatic resources, aesthetics, and recreation.

The most recent OCS orders for the Gulf of Mexico region went into effect October 30, 1970 and were certainly influenced by the Chevron oil spill earlier that year. These orders require more safety devices than did previous regulations. Many operators have reported difficulties and high costs in conforming with these orders (7), believing them to represent a new USGS requirement that no oil pollute the sea in any event. Actually, a pollution prohibition has been in the Secretary's regulations for many years. Even prior to the Santa Barbara incident, the Secretary's regulation with regard to pollution stated, "The lessee shall not pollute the waters of the high seas or damage the aquatic life of the sea or allow extraneous matter to enter and damage any mineral or water bearing formation. The lessee shall dispose of all useless liquid products of the well in a manner acceptable to the supervisors." Following the Santa Barbara accident, this regulation was rewritten in much greater detail, but with no changes in intent. Thus, the philosophy of the USGS was not changed by the new orders; rather, it was the enforcement of this philosophy that was changed.

Safety Devices in Drilling and Operation of Offshore Wells

Many safety devices and procedures are used to minimize pollution risks in the drilling and operation of oil wells. Casings made of heavy steel pipe are implanted in the walls of the well to support them. The well is temporarily filled with "mud," which cools the drill and circulates to carry drill cuttings out of the well. In normal drilling operations the pressure of the mud exceeds the pressure in the oil and the oil is contained. If the oil should begin to escape out the top of the well, the opening between the casing and the drill can be closed by a system of valves called blowout preventers.

Before the well starts producing oil, safety valves are installed at various points so that the well can be closed off at these points if necessary.

Use of Casing

Casings are used to support the sides of oil wells and to support the geological structure in which oil drilling and production take place. Federal requirements for casings are set out in OCS orders for the Gulf and Pacific regions. A number of questions have arisen about the casing used in the region of the Santa Barbara accident (10). A summary of the requirements of OCS order No. 2 for the Pacific region at the time of this accident is taken from Congressional hearings on the Santa Barbara spill (10):

> Conductor casing (the several casing strings in sequence of installation are conductor, surface, intermediate and production) set at a minimum of 300 feet and a maximum of 500 feet below ocean floor unless specific justification is made for approval prior to drilling below 500 feet. Conductor casings cemented in place along its entire length to the level of the ocean floor. Surface casings set at depths not exceeding 3,500 feet below ocean floor. Minimum depths below ocean floor are

Proposed well depth	*Minimum surface casing*
to 7,000 feet	25% of proposed total depth of well
7,000 to 10,000 feet	2,000 feet
below 10,000 feet	2,500 feet

> Surface casing cemented in place along its entire length to the level of the ocean floor. Intermediate casing used as required by well conditions. When used, sufficient cement shall be used to isolate all productive zones. Production casing shall be cemented in wells prior to completion as a producer and in a manner to isolate all productive zones.

A variance from these regulations was granted to the Union Oil Company by the USGS prior to the drilling of well No. A-21 in the Dos Cuadras offshore oil field in the Santa Barbara channel. Further information about this accident can be found in "Geology, Petroleum Development, and Seismicity of the Santa Barbara Channel Region, California" (51). This report states that at the time of the blowout the well was drilled to a depth of 3,313 feet and conductor casing was cemented to a depth of 238 feet below the sea floor. Presumably, the variance allowed the conductor casing to be placed at so shallow a depth in spite of the OCS order that requires a minimum conductor casing depth of 300 feet below the sea floor.

One surmises that shallow casing depths were desired by Union Oil because petroleum deposits existed at very shallow depths in the Santa Barbara Channel. Congressional testimony (10) indicates that petroleum-saturated sand has been found as high as thirty feet below the sea floor. These hearings also offer an explanation for the sequence of events that occurred during the blowout. When the well operators closed off the well at the top, fluid pressure built up in the well and blew out just below the bottom of the casing. Here the use of the deeper casing required by the regulations might have prevented the accident. The hearing transcript (10) describes the very complicated geology of the bottom of the Santa Barbara Channel. This complexity seems to indicate that a more thorough understanding of the geology in drilling regions and the imposition of casing requirements consistent with this geology would effect the best compromise between economy and safety.

Drilling Mud

Drilling mud, whose primary use is to remove rock chips cut by the drill, serves the important secondary purpose of a defense against blowouts (47). Drilling mud is clay that has been dried, purified, and rehydrated with salt or fresh water. Other substances, such as starches and barium sulphate, are sometimes added to drilling mud to tailor its properties to the requirements of a particular well. The mud must be kept heavy enough to contain pressures encountered during drilling, but not so heavy that it fractures the earthen strata around the well. Thus, it is most important to control the properties of the mud. Apparently experience plays a large role in deciding what is best at a given time, and thus determining optimal mud properties is something of an art.

Blowout Preventers

Blowout preventers are required by law when drilling goes below the conductor casing. A blowout preventer is a device located near the top of the string of tubing through which a well is drilled. Most wells have three preventers. The lower two preventers are ram-type preventers, with two steel rams that come together across the tubing. The lower preventer usually has blind rams, which close the hole completely when mated. The preventer above it is usually made with cutouts in the rams so that the preventer can close around the drill shaft. The uppermost preventer is usually a bag-type preventer. It contains a bag shaped like a doughnut and is inflated with fluid to close the hole, which may be empty or have a drill shaft in it. The bag-type preventer, although somewhat less secure than the ram-type preventer, allows the drill shaft to move with the preventer closed. Although preventers are principally used during drilling, they are also frequently used to prevent accidents when the production tubing is removed from the well (8).

Blowout preventers were not in use in the Shell Oil well blowout of the winter of 1970-71 (48). Apparently the well had become clogged with pieces of plastic from the lining of a pipe, and a contract crew had been hired to unclog the well. There was no subsurface safety valve in the well, as USGS had granted a waiver on the requirement for the safety valve. The work crew had inserted a wireline–a long taut wire–into the well for purposes of controlling the unclogging equipment. At one point the crew left the well unattended and the control valves open. It was then that the blowout and fire occurred, killing four people. The Shell Oil report of the incident (48) makes no mention of blowout preventers. However, Shell's manager of environmental affairs has claimed in writing that a blowout preventer was being installed at the time of the accident (56).

In the Santa Barbara blowout, closing the blowout preventers resulted in oil coming out through the fractured seabed outside the casing. Thus, in this case, the action intended to prevent pollution worsened the situation. This exemplifies the need for both operators and regulatory agencies to base decisions primarily on an understanding of specific situations rather than to rely entirely on general operating regulations.

Surface Safety Valves

The group of pipe valves and gauges on the top of an oil well is referred to as the "Christmas tree" (Figure 2-7). In most wells, the oil passes through the horizontal arm of the "Christmas tree," and the flow can be shut off manually with a wing valve. Farther away from the well, a surface safety valve forms the first line of defense against potential polluting accidents above the well by shutting when any of the fusable plugs on the platform melt in the event of fire.

Erosion is a problem encountered with most oil well safety devices. Frequently, a substantial amount of sand is produced by the well, and the sand tends to erode the parts of the flow system. However, erosion can occur even in the absence of sand by the moving fluid itself; it is often related to unsteadiness in the flow when a well is simultaneously producing oil and gas. An erosion detector (9) has now been developed to close the surface safety valve if erosion damage potential becomes large. This detector is a small tube, sealed at one end, inserted in the production tubing next to the surface safety valve. The walls of this tube will be eroded away before serious erosion damage takes place elsewhere in the well. The tube is connected to a pressure sensing device so that, if it should be eroded through, well pressure will be applied to the device, which will close the surface safety valve.

Down-Well Safety Valves

Safety valves in deep oil well production tubing are capable of shutting off the oil flow. They are required unless a waiver is obtained from the

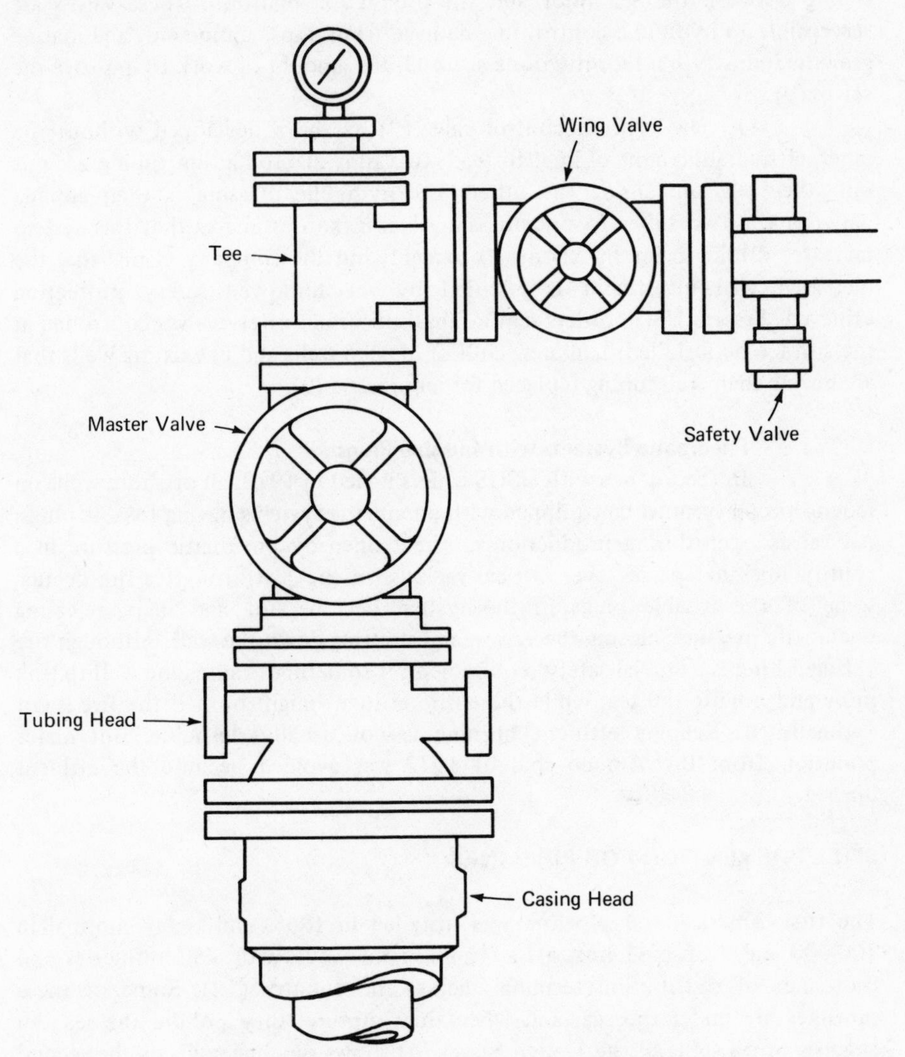

Source: Exxon Corporation, "Preventing Oil Spills."

Figure 2-7. Sketch of a "Christmas Tree".

USGS. One type, usually called a storm choke, is designed to close if the oil flow rate through it exceeds some specified value. Storm chokes have been found especially susceptible to erosion damage, which prevents the valve from shutting down the well if the flow rate becomes excessive. A second type of valve is activated by hydraulic pressure in a hydraulic control line strung external to the tubing between the sea floor and the production platform. These valves are susceptible to hydraulic control line damage from ships, equipment, and marine growth. Industry has recently done a substantial amount of work to improve the valves (9).

A new type of control valve (9) has been developed without the external hydraulic control line. Instead, two runs of production tubing are put into the well, one inside the other. The hydraulic pressure is then applied between the two tubes to operate the valve. Exxon estimates that this system increases drilling costs by about 10 percent; but the company claims that the increased well investment is justified by the improved safety protection achieved. Present USGS orders require that subsurface safety valves controlled at the surface be installed in all new Gulf of Mexico wells and in existing wells that are having their well tubing replaced for any reason (9).

Pneumatic Systems with Fusable Plugs

In accordance with USGS orders issued in 1970, all offshore wells on federal property must be equipped with pneumatic systems having fusable plugs. All valves open during production are held open by pneumatic pressure in a control line that passes over critical regions on the platform. If a fire occurs, some of the fusable plugs in the system become hot and melt, releasing pneumatic pressure, closing the valves, and shutting down the well. Although fire extinguishing is a crucial safety requirement, it sometimes causes the well to leak more and pollute the sea, while the entire effluent might burn if the fire is not extinguished. Keeping effluent burning is not usually desirable. But major pollution from the Amoco spill of 1972 was avoided because the effluent burned.

SPILL PREVENTION FOR PIPELINES

The first American oil pipeline was installed in 1865, and today more than 200,000 miles of pipe link a half million oil wells with 250 refineries and thousands of distribution terminals across the country (12). Some of these pipelines are under the sea, and when they rupture, they pollute the sea. An analysis of oil spills in the United States (4) shows pipeline spills as the second largest source of ocean pollution in coastal waters. There were 1,436 pipeline breaks, with a total spillage of 897,685 gallons in 1971. A breakdown of all pipeline spills in the Georges Bank study (4) shows that approximately 90 percent of the offshore pipeline spills and 97 percent of the oil spilled comes from pipelines leading to wells less than three miles offshore. The pipelines near the shore generally serve the oldest platforms, which are not on federal lands. Most pipeline spillage is the result of failure of the older pipelines. Turner (12) has written that pipe corrosion is the principal cause of pollution from pipelines. Corrosion results from material outside the pipeline, and not from the oil inside the pipeline.

Newer pipelines usually have protective coatings or cathodic protection to reduce the corrosion rate. Many new pipelines also have shutdown devices to stop the oil flow if a major leak occurs.

Frequent inspection of the pipelines provides a major safeguard against pollution. Inspection of undersea pipelines must be made by divers which makes it difficult to be thorough.

Chapter Three

Cleaning Up

The containment, removal, and cleanup of spilled oil are among the most difficult and most misunderstood problems in ocean engineering. The present difficulties are both technological (because physical and chemical understanding of some of the phenomena is incomplete) and sociological (because many untrained people wrongly believe that the problem is simple enough to be solved in short order with present-day technology). Except in a few highly specialized areas–offshore oil well drilling among them–business and government have devoted far less capital expenditure to ocean engineering than to other fields of engineering. For this reason, ocean engineering is a backward field in the sense that many potential areas for technological development have not been pursued to the extent possible. The containment, removal, and cleanup of spilled oil is one such area. The application of modern technology to this problem did not begin on any large scale until the aftermath of the *Torrey Canyon* disaster in 1967.

There are many reasons why oil spill cleanup problems are so difficult. There is a lack of understanding of the physics and chemistry underlying some of the pollution control difficulties. Some oil slicks cover tens of square miles. Currents and waves generate enormous forces on equipment. The logistics of dealing with something so large and so mobile in the face of the large forces of the sea are staggering. The area of the earth susceptible to an oil spill is large, and spills occur at random. To understand the difficulty of dealing with such a problem, one can compare a spill that happens unexpectedly a hundred miles offshore during a severe storm with the launching of a space vehicle from a known position under laboratory conditions following a predetermined countdown procedure in ideal weather. If a storm should occur, the launching of a space vehicle can be delayed, but for a ship it may be precisely such a storm that causes a polluting accident. Asked for his honest recommendation for dealing with a large spill at sea, a former tanker captain from one oil company said, "The best thing you can do is uncork another bottle of whiskey!"

TREATING AGENTS

A number of materials have been used to remove or reduce oil slicks. Treating agents have been used to deal with accidents and field and laboratory experiments have been done to assess their usefulness and to develop the technology to use them. The types of agents which have been used to date are:

1. Dispersants–chemicals forming oil-in-water suspensions;
2. Sinking agents–materials that mix with the oil and create a mixture dense enough to sink;
3. Burning agents–material put on the slick to assist ignition or enhance combustion of spilled oil;
4. Biodegradants–substances that promote oxidation of oil by microbial action;
5. Gelling agents–chemicals that form semi-solid oil agglomerates to facilitate removal;
6. Herding agents–chemicals that concentrate the spilled oil in a small area;
7. Sorbents–materials that absorb or adsorb oil to form a floating mass for later collection and removal.

Dispersants

Dispersants are chemical compounds that act to enhance the surface spreading of oil slicks and to emulsify the oil into the water beneath it. These effects increase the surface area of the slick so that more of it is susceptible to biodegradation. When emulsification (breaking the oil into very small droplets that become mixed with the water) occurs, the slick disperses vertically as well as horizontally. Toxicity of dispersed slicks is a major problem (2,19) and is due both to the effect of the dispersant and to the increased oil surface to which marine organisms are exposed. Considerable research has been devoted to the problem of dispersant toxicity.

The federal Council on Environmental Quality (CEQ) developed a *National Oil and Hazardous Materials Pollution Contingency Plan* in August 1971 (2). This plan, enforced by the U.S. Coast Guard and the Environmental Protection Agency, requires that dispersants not be used:

1. On any distillate fuel oil.
2. On any spill of less than 200 barrels.
3. On any shoreline.
4. In waters less than 500 feet deep.
5. In waters containing major fish populations, or breeding or passage areas for

species of fish or marine life that may be damaged or become less marketable by exposure to dispersant or dispersed oil.
6. In waters where winds or currents could carry dispersed oil to shore within twenty-four hours (in the judgment of EPA).
7. In waters where the surface water supply would be affected.

Dispersants may be used, however, in any place and at any time if, in the judgment of the Coast Guard on-scene Commander, their use will prevent fire or substantially reduce the hazard of fire to property. They may also be used if, in the judgment of the EPA, they will prevent or substantially reduce the hazard to vulnerable waterfowl and cause the least overall environmental damage.

These restrictions do not exist in all nations. For example, in the United Kingdom the use of dispersants is a standard practice for handling oil spills.

Dispersants are of major interest to oil companies because they are relatively easy to use, and for this reason substantial research into dispersants has been supported by the oil companies (20). A Texas A & M University study of superports (21) contains a cost-effectiveness analysis of materials and technology for removal and dispersal of spilled oil. The study concluded that dispersal was the most practical means of dealing with massive spills from a cost-effective standpoint.

The use of chemical dispersants presents two major problems. The first is to obtain adequate mixing between the dispersant and the oil slick; and the second is to minimize toxicity to marine life. The mixing difficulty was demonstrated in the Santa Barbara accident, where it was found that after the dispersant was spread in a fine mist over the oil slick, it was necessary to run work boats through the slick so that their propeller action would mix the dispersant with the oil (11). Work is currently underway on the development of dispersants that will require little or no mixing. However, the question of toxicity remains. Some dispersants are more toxic to marine life than others, and the differences may relate in part to varying quantities of oil surface to which organisms are exposed. As oil slicks are dispersed into droplets, the surface area multiplies.

A less serious problem with dispersants involves the stability of the oil-water emulsion they generate. Experiments have shown that with some dispersants the lifetime of this emulsion can be relatively short, and the dispersed oil soon recombines into a slick. However, the stability of the emulsion generated by other dispersants is quite long (11), and a number of companies are doing work on increasing the emulsion stability.

Dispersants have been found effective when it is advantageous to disperse small parts of an oil slick. For example, dispersants were used and found effective to remove slicks in the vicinity of the burning oil platform at the

Chevron Oil spill (6). However, it would have been impossible, even if allowed by law, to completely disperse all the polluting oil from this incident. If dispersants requiring no mixing are developed in the future, it may become possible to disperse an entire slick of moderate size.

Sinking Agents

Dense materials that can be mixed with an oil slick may enable the oil to sink. Present law allows the use of sinking agents only under special circumstances and with the approval of the Environmental Protection Agency.

Following the sinking of the *Torrey Canyon*, the French sank a large amount of its oil off the coast of Brittany by spraying hydrophobic chalk on the surface of the sea (22,23). The literature on the *Torrey Canyon* states that this chalk floated on the surface of the sea until it came into contact with floating oil, causing the oil to sink after a few hours. It has been estimated that 20,000 tons of oil were sunk with three thousand tons of powdered chalk. Very little of the sunken oil came ashore, and apparently no diving expeditions were made to examine the oil on the bottom of the sea or its effect on bottom life. However, it has been reported that no oil was later found on the nets of fishermen, and fishing was not adversely affected (23). All the sunken oil seemed to disappear from normal sea operations (22). In light of the apparently outstanding success of sinking oil with chalk off Brittany, surprisingly little information is available about this method. It is somewhat disconcerting to note that one experimental study on sinking agents (23) does not even mention chalk, although it states that the *Torrey Canyon* accident motivated the study itself.

A test of an attempt to sink oil with talc after the Santa Barbara incident (11) was found to be totally unsuccessful. The safety and toxicity of sunken oil are unknown; moreover, the difficulty of determining them has apparently discouraged most research into finding new sinking agents.

Burning Agents

Burning agents are materials or chemicals to aid the burning of oil and water. Different burning agents work in different ways, although some agents combine various ways. Some materials act as hydro-igniters, which react violently with water to produce hydrogen and heat with explosive violence. Other materials act as auto-igniters, which react with the combustible components of oil to form a semi-explosive mixture.

Other burning agents, such as lightweight oils, are ignition assistors, which can burn easily and maintain temperature above the flashpoint of a slick of heavier oil such as crude. Still other materials are wicking agents, which draw up the oil into them, provide an increased oil surface area adjacent to air, and thermally insulate the burning oil from the surface of the sea as well.

Some research has been done on the use of burning agents (49). The basic conclusion from this work is that reasonably small and relatively thick slicks of oil can be burned with these agents. Smallness of slick is required because of the necessity of seeding the slick with burning agents.

One of the major difficulties in the use of burning agents is incomplete combustion of the oil, which results in a substantial amount of air pollution and the manufacture of residues that may diminish combustion, sometimes to the point of extinguishing it. Some work has been done to develop combustion agents that would burn a pool of oil completely. However, at the present time, no effective burners exist for use on the sea. It may be possible to burn a slick by using a towed barrier to keep the slick thick near the barrier, where a burning agent is applied.

There are a number of other practical problems involved in the use of burning agents. One burning agent, a silane-treated fumed silica, is designed to prevent water from sticking to the small pieces of silica. This material is very light in weight. Although it can burn oil heavier than itself, its volume must be approximately equal to that of oil to be burned. Thus, to burn a supertanker full of oil, a supertanker full of treated silica would be needed. Expediting such a large volume of material to the scene of an accident poses practical difficulties.

Biodegradants

When oil is spilled on the sea, some of its lighter parts evaporate, but the ultimate fate of the remaining oil is biodegradation of the hydrocarbons by marine micro-organisms. It has been found that the natural marine environment contains certain bacteria capable of degrading oil, even at relatively low temperatures (25). The idea of enriching oil hydrocarbons by using bacteria or yeast has been advanced by biologists for quite some time, but the literature shows no quantitative data about the effectiveness of this technique. However, oil industry representatives report that a large number of relatively small oil spills disappear by themselves with no special cleanup effort, especially in heavy seas and warm weather. This suggests that it is indeed possible for biodegradation to occur relatively quickly under some conditions. The concept of enhancing biodegradation of oil spills is worthy of more study.

Gelling Agents

When a tanker is damaged, and removal of its oil through a transfer operation proves unfeasible, pollution can sometimes be prevented or diminished if the oil in the tanks of the ship can be gelled into a rigid, solid mass. Such a gel could remain in the tanker even if the tanker sank or were damaged further, under certain conditions. If it becomes feasible to transfer the oil off the tanker at a later time, most oil gels can be reliquefied by heating them to a temperature of about 130°F. Some experimental work has been done on the gelling of oil in tankers (50).

Gelling oil requires adequate mixing of the gelling agent with the oil and considerable time for the gel to set. Mixing studies (50) indicate that a jet nozzle mixer moved about through the tank is an adequate way of mixing, although such devices are not routinely built into tankers. Approximately eight hours is generally required to achieve a gel of modest strength; the gel doubles in strength after about thirty hours and triples in strength after 130 hours. No further gain in strength seems to occur after 130 hours. Goldstein *et al.* (50) estimate the cost of gelling oil at approximately $4.50 per barrel. On a tanker the size of the *Torrey Canyon*–modest by comparison with many built in the last few years–the total cost would be $2.7 million.

Herding Agents

A herding agent alters the surface tension of oil-water, water-air, and oil-air interfaces so as to contract an oil slick rather than spread it. When a slick attains any modest thickness, gravitational effects result in spreading forces. Thus herding agents can thicken an oil film only to a limited extent. As a result, herding agents are only useful for thin slicks containing small amounts of oil.

Sorbents

Sorbents may either absorb or adsorb oil to form a floating mass for later collection and removal. Sorbents, used either with or without a containment barrier, are considered by the author to be the only treating agent that can be used safely and effectively at this time. Future developments are expected to make them much more effective, and their potential is a fruitful area for research.

Historically, the most common method of dealing with relatively small oil spills has been to use straw as a sorbent. Straw can absorb approximately five times its own weight in oil. After the straw absorbs as much oil as possible, it is gathered together and disposed of, frequently by burning after evaporation of the water. As a result of a substantial amount of research on sorbents (26,27,28,29, and 30), the best sorbent material appears to be reticulated polyurethane foam. Not only does it absorb about thirty times its own weight in oil, but the oil can be largely removed by passing the material through wringers, so that the reticulated foam can be rebroadcast onto the sea to absorb more oil. Although no full-scale continuous-use sorbent equipment has been built and used on the sea, Miller (29) estimates that the cost of a unit designed for operation in protected waters, with a recovery rate of three thousand gallons of oil per hour, would be $76,480. The three thousand gallon-per-hour recovery rate is what investigators expect of future sorbent systems. It is considered to be quite slow for dealing with large spills at sea, but adequate for many small spills in harbors.

The effectiveness of a sorbent system is not seriously diminished by adverse sea conditions. In fact, better results can sometimes be obtained in the presence of waves than in calm water. If a mechanical sorbent retrieval system is not used, manual labor is required for recovering the sorbent, and this is a major disadvantage. The possibility of using a sorbent together with a pollution control barrier is interesting to researchers, although none has investigated the details of how such a system could work. Combining a sorbent with scattering equipment, a pollution control barrier, and an oil removal system offers a possibility for future oil pollution cleanup. Although the expense for a system large enough to work effectively on large spills at sea seems great—several hundred thousand dollars—this cost is small compared with the potential damage costs from a large spill.

OIL POLLUTION CONTROL BARRIERS

An oil pollution control barrier, frequently called an oil boom, is a device floated on the surface of the sea to prevent the passage of an oil slick from one side of the barrier to the other. A pollution control barrier looks like a vertical curtain piercing the surface of the sea to a depth greater than the thickness of the oil slick. To be effective, the barrier must follow the motions of the waves so that its top never goes beneath the top of the slick and its bottom never rises above the bottom of the slick. Typical barriers have a vertical height varying from six inches to five feet, with between 55 percent and 90 percent of this vertical height below the sea surface and the remainder above the surface in calm water (see Figure 3-1). For many years pollution control barriers have been used to contain oil from leaking ships in harbors where the water is typically very calm. There it is relatively simple to surround the slick with a curtain ballasted at its bottom and buoyed up near its top to keep it vertical and contain the oil between a ship and barrier. This can be done with a fairly simple barrier. However, when taken away from protected harbors into moderate currents and waves, barriers designed for calm water break apart easily and are unable to contain oil even if they do not break. The problems of using a barrier in an unprotected area are caused by currents and waves.

In a current, a barrier is intended to hold the oil against the current. One technological challenge is to make the barrier remain vertical at the right height without rising or sinking. Another is related to the hydrodynamics of containing the oil. Both of these problems have been extensively researched (31), (34), (35) and (52).

Ocean engineers can now design barriers that will remain vertical in moderate currents. Although a number of satisfactory barriers exist, many barriers on the market still fall short. Of the satisfactory calm-water barriers, the best in any circumstance is usually the one that can be deployed soonest. Even with the satisfactory calm-water barriers, containment is restricted by two hydrodynamic effects. First, for any barrier depth in a specific current speed, there is a limit to the amount of oil that can be contained. The limit is passed when the pool of oil held by the barrier is so deep that it flows beneath the

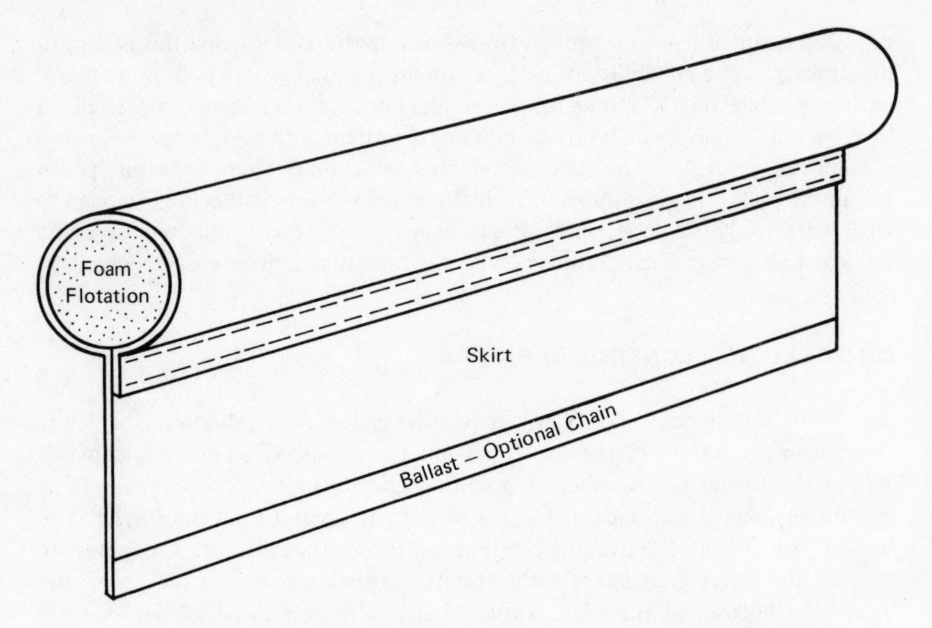

Figure 3-1. The Kepner Barrier–The Model Used by the U.S.C.G. has a Chain in the Bottom of the Skirt.

barrier. An oil slick piles down below the waterline, with only about 10 percent remaining above the waterline. In other words, an oil slick floats in much the same way that an iceberg does. Moreover, there is a critical current speed for each oil density and boom depth above which a barrier will not be able to contain any oil (4). The critical speed for a barrier having a draft of one foot holding oil whose density is 10 percent of water is 0.95 knots. If the current speed exceeds 0.95 knots in such a situation, the barrier will not be able to hold any oil because the oil will be driven against the barrier so hard that it will go down the barrier and pass underneath it. This is called drainage failure. In certain conditions, the critical speed problem can be overcome in part by mooring a barrier at a steep angle to a current (31) and collecting the oil and removing it from the region where it accumulates (see Figure 3-2).

The second problem involved with the hydrodynamics of containing oil is called entrainment failure. When a slick is being held by a barrier against a current, droplets of oil are "entrained" in the moving water. This is a condition in which a dispersion of oil droplets are driven into the water beneath the slick. Some of these droplets move deeply enough to pass underneath the barrier. A substantial number of droplets form at the leading edge of the oil slick far from the barrier (Figure 3-3), where the oil slick has a lump, called a headwave,

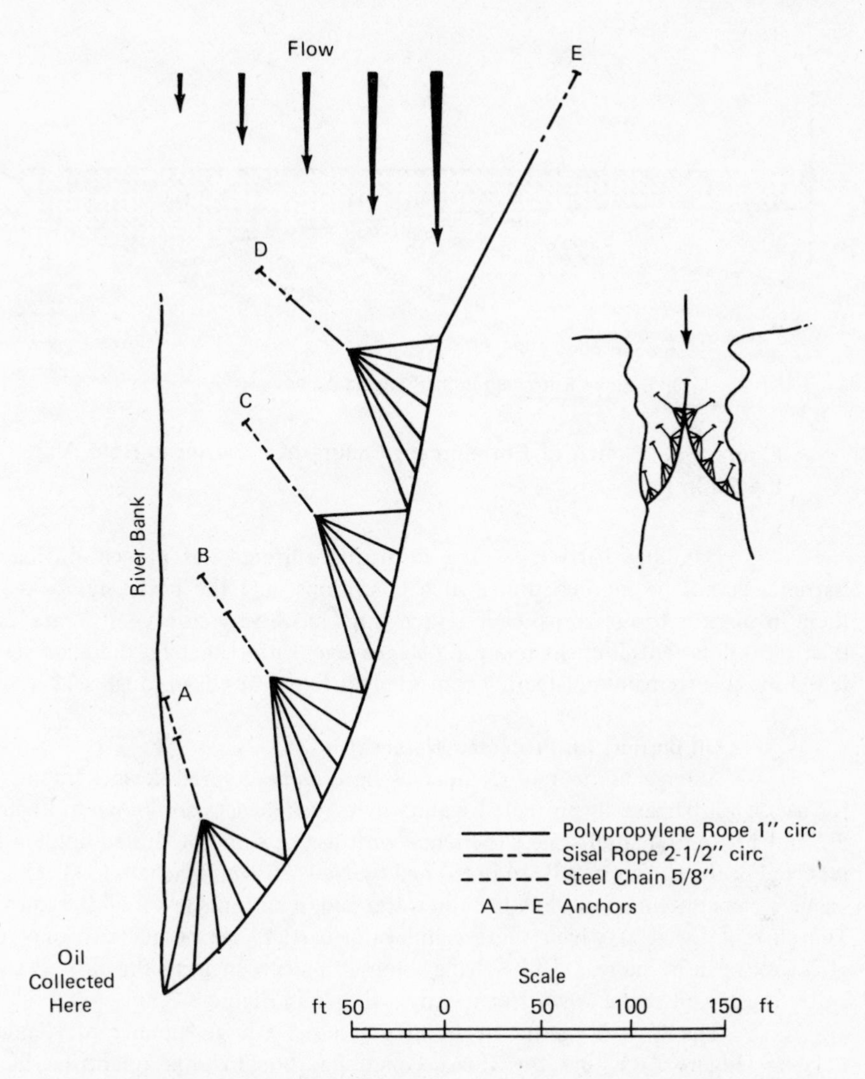

Figure 3-2. Barrier Moored at an Angle to a Current.

intruding down into the water. Further entrainment comes from the main body of the slick between its leading edge and the barrier. The entrainment phenomenon varies widely with current speed. Evidence indicates that entrainment is very small in current speeds below 0.5 knots, and very large at a current speed of 1.2 knots (33,34,52). The properties of the oil itself affect the entrainment rate in ways not completely known. Technologists understand drainage failure well enough to design all equipment necessary to prevent it in given conditions, but this is not true of entrainment failure.

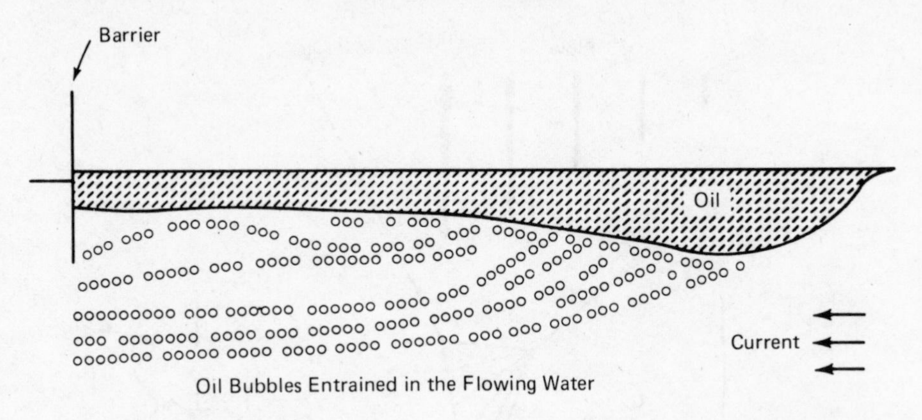

Figure 3-3. Sketch of Entrainment Failure of a Barrier to Hold All the Oil In It.

Handling barriers at sea in strong currents has proven difficult. Barriers cannot be moored under such conditions, and the best way to keep them in place is to use ships. Unless such ships move very slowly—at a rate less than 0.6 knots--entrainment loss will be excessive. Unfortunately, there are very few ships able to maintain steering control at such low speeds in rough seas.

Oil Barriers for Protected Waters

A large number of companies have manufactured devices intended for use as oil barriers in protected waters; two such devices are shown in Figure 3-2 and Figure 3-4. Full-scale experience with use of some of this equipment is reported in the Chevron spill study (6) and by Newman and Macbeth (31). These barriers can contain an oil slick in calm water and in currents less than 0.8 knots. The current speed at which these calm water barriers can be made to operate effectively can be increased by setting them at a steep angle to the current and collecting the oil at the downstream end, as shown in Figure 3-2.

The U.S. Navy has recently purchased a large number of Kepner Barriers (Figure 3-1), and the Coast Guard has bought large quantities of a modified version of them. The modification involves installation of a chain at the bottom of the barrier to serve both as a tension member and as ballast for the barrier. The Coast Guard recently used approximately three thousand feet of this barrier in San Francisco Harbor. Spokesmen from the Ocean Engineering Division of the Coast Guard have reported that, in this case, effective operation could be obtained only when the barrier was set at an angle to the current to divert the oil towards collecting devices. It was also reported by the Coast Guard Ocean Engineering Office that, of the three thousand feet of barrier deployed, about two thousand feet were damaged. This damage occurred in spite of rigid specifications demanded for the physical strength of the barrier prior to its

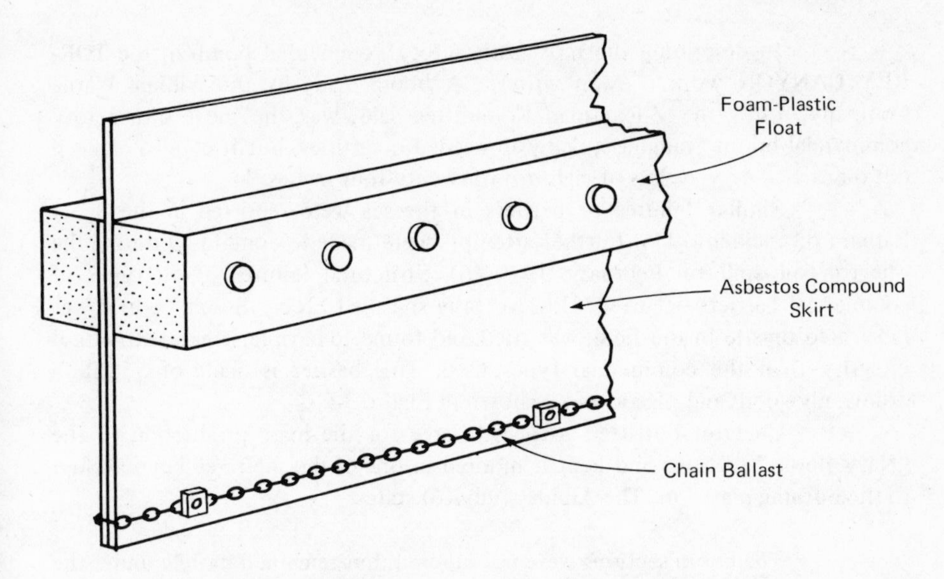

Figure 3-4. Sketch of the Spillguard Barrier. (from reference 6)

purchase. The Coast Guard believes most of the damage was produced by two circumstances not related to the actual containment of oil. First, a large amount of floating debris damaged the barrier. Second, the barrier was also damaged by the method used to handle the barrier during deployment and recovery. For these reasons, the Coast Guard Ocean Engineering Office expects some of its research emphasis to turn away from optimal oil containment performance towards increased ruggedness and ease of handling to avoid damaging barriers.

Oil Barriers in the Open Sea

The first reported attempts to use oil pollution control barriers in the open sea followed the grounding and eventual breakup of the TORREY CANYON in the English Channel (18). These attempts, made with calm-water barriers, all resulted in failure. Cowan (18) describes some of them. One barrier taken to sea was the Aeropreen barrier. In Cowan's words (18), "The Aeropreen Barrier was installed at the entrance of the Fal Estuary, where, in relatively mild conditions, the whole thing came apart." In describing another barrier bought from a firm in the U.S., Cowan says, "Nor did it work well; the oil slipped under it or the waves broke the moorings or the couplings, particularly on the Porthmeor Beach, which is more exposed to the open sea than is Saint Ives Harbor. This was the experience along the Cornish and Devon coasts. Similarly, attempts to gather up oil with booms failed. The oil invariably slipped under the boom's skirt."

In describing the most satisfactory commercial boom in the TORREY CANYON work, Cowan writes, "A boom made by the William Warne Company, Ltd., the Zuckerman Committee said, was the most satisfactory commercial boom available quickly in needed quantities, but it could not keep out oil carried on wavelets of eighteen to twenty-four inches."

Similar failures of barriers in the sea were reported in the Santa Barbara oil incident (11). Further attempts were made to contain oil during the Chevron oil spill of February 1970 (6). Structural failures of all types of commercial barriers occurred. However, the so-called "Navy Boom," which was fabricated on-site in the field, was tried and found to have far greater structural integrity than the commercial types used. This barrier is made of 55-gallon drums, plywood, and plastic and is shown in Figure 3-5.

Chevron outfitted a special barge for the mass production of the "Navy Boom" sections and kept it moored in one of the sheltered bayous close to the burning platform. The Alpine study (6) states:

> The boom sections were not moored, but remained mobile under the control of a pair of tugs. Two 500-foot sections of the "Navy

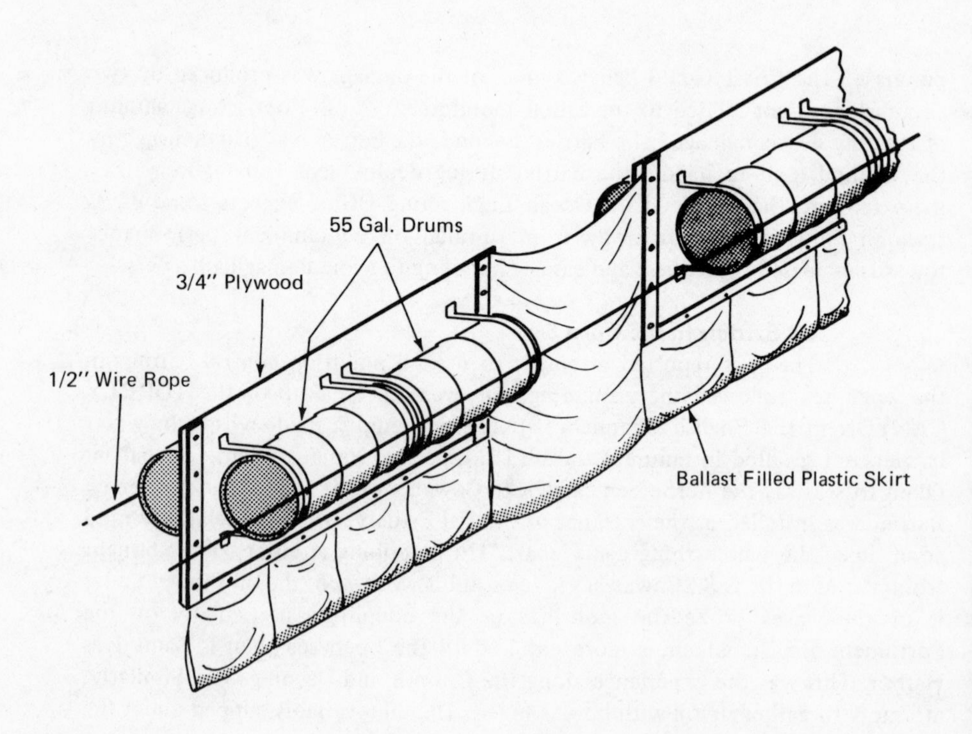

Figure 3-5. The "NAVY BOOM".

> Boom" were combined to make a V-shaped barrier which concentrated the oil for pickup by the skimmer barge and boats. The two tugs were used to spread the mouth of the "V" and the skimmer vessel was held at the apex. It was felt that the boom could not stand the strains which would be exerted if the apex was completely closed off, so a gap was at the apex through which the oil flowed to be caught in the containment booms of the skimmer vessels and pumped aboard. . . . For this kind of operation, the "Navy Boom" maintained its integrity in up to six-foot seas.

Although the "Navy Boom" was structurally adequate for seas up to six feet high, it could not contain oil in waves as high as four feet.

In order to develop a barrier having adequate structural integrity and to contain oil in moderate conditions, the Coast Guard research and development office has undertaken a major program in high-seas oil-barrier development (32,33,34 and 35). To date, approximately 2,860 feet of full-scale prototype barriers have been constructed by the Johns-Manville Products Corporation under contract to the Coast Guard (35). This barrier has been tested at sea on numerous occasions, twice with full-scale experiments in the containment of soybean oil. Soybean oil is rapidly biodegraded and nonpoisonous, and was used to avoid the environmental difficulties presented by petroleum oil. In the first instance, 27,000 gallons of soybean oil were spilled onto the sea and contained by the barrier, which was towed by two vessels. Towing lasted about eight hours. It was found that at a tow speed of 0.5 knots there was very little oil lost due to entrainment, and that at a tow speed of one knot, the rate of loss was substantially greater. Following one of these towing tests, the oil was recovered with skimming devices placed in the oil pool and attached to an oil-holding barge by hoses; 18,000 gallons were recovered, nine thousand gallons having previously been lost by entrainment. The second time the soybean oil was spilled, the barrier tow speed was purposely increased until the oil was lost in order to be able to evaluate the entrainment effects (34). Structural integrity of this barrier was preserved in waves as high as twenty feet and in currents as swift as seven knots. It is likely that the barrier could withstand even more severe conditions, although it was not tested to the point of structural failure.

The Coast Guard-Johns Manville barrier is shown in Figure 3-6. It was designed with flotation elements that are air-inflatable in order to minimize the packing space. The Coast Guard hopes to have the barrier packaged in an air-deliverable container to be parachuted to the site of an oil spill and deployed there.

Some oil companies have undertaken programs for development of high seas barriers (36). However, no quantitative data about the effectiveness of these barriers is yet available.

Pneumatic Oil Pollution Control Barriers

If air is pumped into a perforated pipe placed below the surface of the sea, a rising curtain of air bubbles results. This rising curtain entrains water

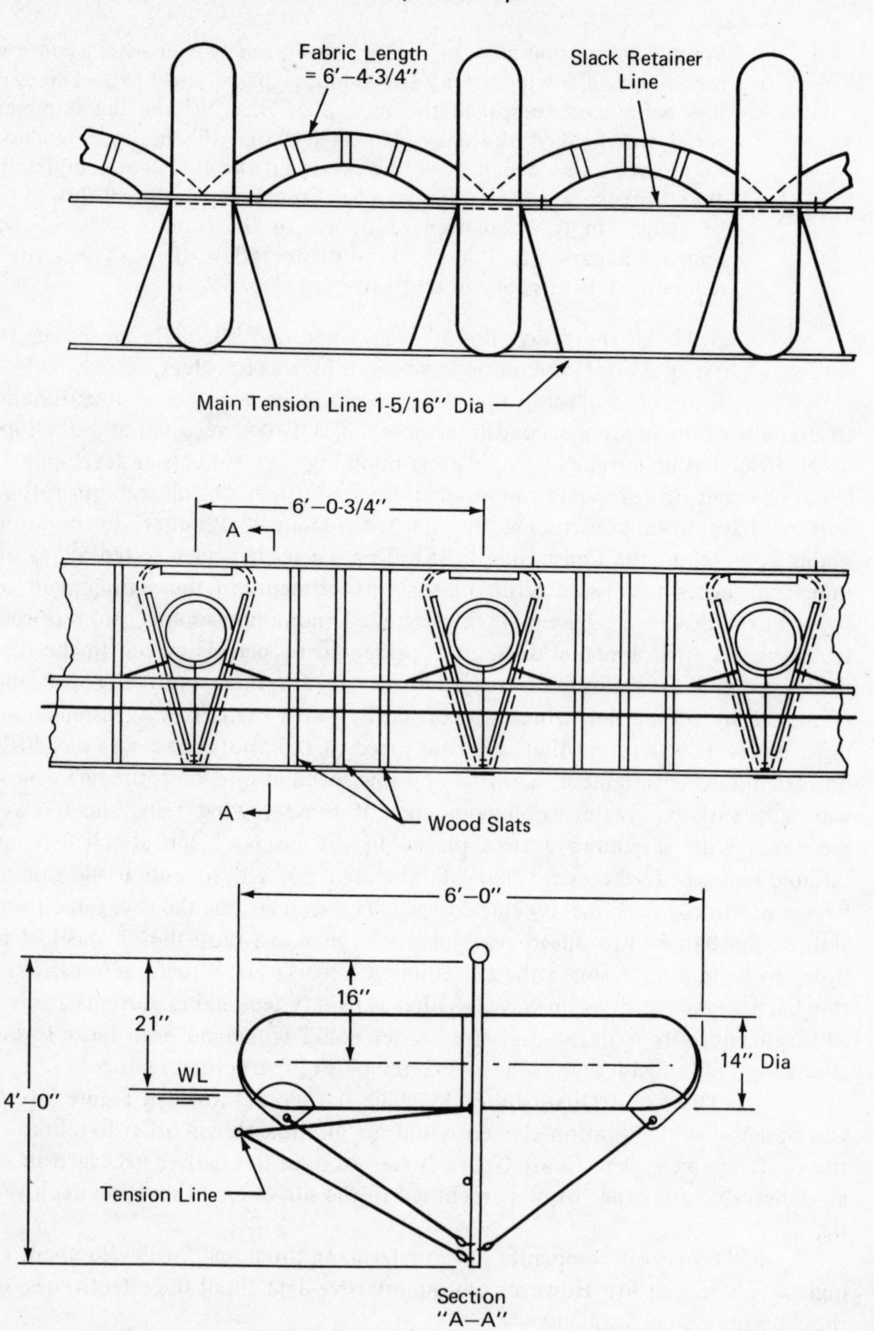

Figure 3-6. The USCG Johns-Manville High Seas Oil Pollution Control Barrier. (generation 1)

with it, making a line of upwelling water. When the water reaches the surface, it spreads horizontally in the two directions perpendicular to the submerged pipe. Experiments indicate that it is possible to stop oil from crossing this two-way current, which has been done for some time in European harbors built especially for the unloading of oil tankers. There, the places where pollution must be stopped are quite narrow and the water especially calm. The air-curtain concept is particularly useful since it can be installed across the mouth of a harbor and still allow ships to pass it. Considerable laboratory research has been done to determine requirements for air barriers (37). One conclusion from this work is that fairly large amounts of compressed air are needed, typically one horsepower per foot of barrier. An air-curtain barrier was used at the Santa Barbara oil spill (11). It was not successful owing to operational problems, and thus the effectiveness of using an air barrier at sea could not be determined conclusively.

OIL-SKIMMING DEVICES

In addition to oil pollution booms, numerous companies have designed and built devices for skimming oil from the surface of the sea without first doing the necessary background work. Although some of them are effective for very small harbor spills, their use at sea was a failure. One important thing that did not receive sufficient consideration for years is that, for effective oil pickup, a large amount of sea surface must be swept; however, the sweeping speed must be kept moderately slow to keep large amounts of oil from going past the skimming device. It has been conclusively shown that the fluid velocity into any type of oil-pickup device is limited by the depth of the oil (35,53): the deeper the pool of oil at the pickup device, the higher the allowable velocity. Once this limiting velocity is reached, a greater velocity will not increase oil pickup, but merely pick up more air and water. Therefore, in order to be able to pick up a substantial amount of oil, a fairly deep pool at the pickup device is needed. These facts indicate that the most effective type of pickup device is a boom having a collector inside it, with the oil driven into the boom either by current or by towing the boom. The boom provides a large surface area from which to collect the oil into a relatively deep pool at a skimmer, which actually carries out the collection. The collection of 18,000 gallons of soybean oil (35) was done like this, as was most of the collection carried out during the Chevron oil spill (6).

Although an oil collection device covering a large area is required for large spills at sea, there are some instances where the skimming of smaller slicks would be facilitated by a collection device that could be effective without a boom.

Suction-Type Skimmers

A suction-type skimmer is a device that floats on the sea and has openings near the waterline. Suction pumps are hooked to the device to draw in fluid through these openings. In actual use, they collect both oil and water. To be able to work effectively in waves, the motion of the oil collector must follow that of the sea surface quite closely. Numerous suction-type skimmers have been described (35,36). Because such skimmers are relatively small, effective use of them on sizeable spills can only be made in conjunction with a barrier. In recent verbal communications with the Ocean Engineering Office of the U.S. Coast Guard, it was found that suction skimmers were effective in an oil spill with an oil barrier deployed at a steep angle to the current and the barrier acting as a diverter and moving the oil towards the skimmers.

Although use of suction-type skimmers has only been reported for relatively calm conditions, model tests have shown that some of them can be effective in rough conditions. This indicates that it is possible to design and construct suction-type skimmers to follow the surface of the sea well enough to work in large waves. However, only one test of open-sea operations combining suction type skimmers and booms has been quantitatively reported (35). The skimmers used in that test were each designed for a maximum flow rate of 13,000 gallons per hour. This flow rate is too small for picking up massive ocean spills when the oil slick is thick, but adequate for sweeping operations in thin slicks. For thick slicks a number of skimmers can be operated with each barrier.

Weir-Type Skimmers

A weir-type skimmer is designed for the oil to flow over the "weir"–a vertical dam with its top slightly below the surface of the sea or oil slick. After the oil flows over the weir, it passes into a recovery device, frequently through a suction pump. Glaeser (36) describes development of one weir skimmer, called the Exxon Skimmer, shown in Figure 3-7. The constraints on operating conventional weir skimmers are the same as those for suction-type skimmers. The oil slick must be thickened by use of a barrier in conjunction with the skimmer in order to achieve a high collection rate. A special type of weir skimmer has been developed by Ocean Systems, Inc., under contract to the Research and Development Office of the U.S. Coast Guard (38). This is actually a weir skimmer integrated with an oil barrier. Other barriers may be attached to each side to form a large barrier with the skimmer in its center so that the whole configuration can be slowly towed through an oil slick, as shown in Figure 3-8. The weir over which the oil first flows in this skimmer produces a relatively calm pool of oil behind it. Initial tests of this skimmer at sea indicated it was too weak to remain intact in seas over six feet high. However, its concept appears viable for collecting very large amounts of oil from a single barrier. If it were developed into a manageable, structurally sound device, it would be useful for collecting thick oil slicks.

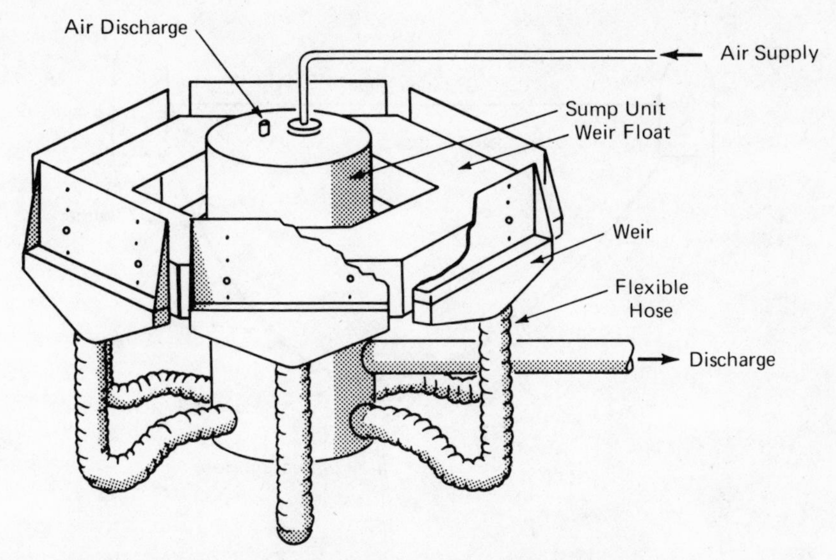

Source: Glaeser, "An Effective Oil Spill Containment–Recovery System for High Seas Use," Proceedings of Joint Conference on Prevention and Control of Oil Spills, Washington, D.C., 1973.

Figure 3-7. The Exxon Weir-Type Skimmer.

Surface-Type Skimmers

A surface-style skimmer moves a solid surface or oleophilic belt across an oil slick and absorbs some of the oil. An oleophilic material is one that can absorb a lot of oil, and it generally absorbs more oil than water. The oil that has adhered to or has been absorbed into the moving surface is then scraped off or squeezed out into a collector before the surface is again put into contact with the oil slick. The earliest surface-style skimmers were belts or drums that were rotated through the sea surface. The adhering oil was then scraped off. The principal difficulty with these devices was that the collection rate was far too low. An improved version has been developed by the Lockheed Company under contract to the Coast Guard (39). The surface area of the material is in contact with the sea, and the rate of oil pickup is greatly increased in the Lockheed device over that of conventional belt or drum skimmers. This is accomplished by using a row of discs at the surface to pick up the oil; in fact, the whole device looks very much like a farmer's disc harrow. It shows promise, although designs of this type have a tendency to drive oil away from themselves because of the relative motion between the device and the oil and water.

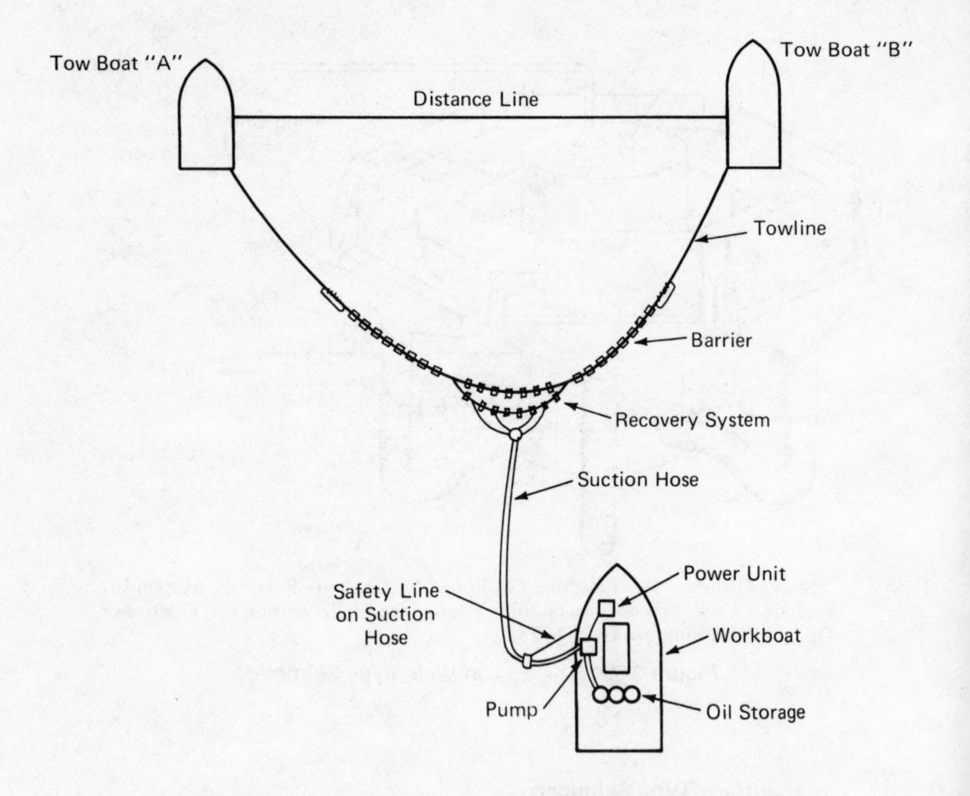

Source: March and Beach, "High Seas Oil Recovery System," Proceedings of Joint Conference on Prevention and Control of Oil Spills, Washington, D.C., 1973.

Figure 3-8. Integral Barrier and Skimmer for Collecting an Oil Slick.

Chapter Four

Critical Areas in Need of Research

Practicality must be a prime consideration for oil spill technology. Although it would be possible to design a strong pollution control barrier fifty feet deep with a 30-foot sail above the waterline, the cost and effort involved would make it impractical. Only reasonably practical areas are considered here. Nevertheless, since improved spill prevention and control will be expensive, one can expect oil industry resistance to some of the research areas suggested here.

RESEARCH ON SPILL PREVENTION FROM WELLS

The research most urgently needed to improve the prevention of spills from oil wells concerns operating practices. These practices include both those of the well operators and those of the USGS. The facts indicate that it was inappropriate for USGS to grant variances for the casing depth at the site of the Santa Barbara blowout and on well safety devices at the site of the Shell Oil well blowout in the Gulf of Mexico. However, in each case, there was no objective way of knowing prior to the accident whether a variance was appropriate; the variance was granted in a subjective decision by a USGS supervisor.

It would be to no one's benefit if USGS were merely to make variances more difficult to get by involving more levels of bureaucracy. What is needed is an objective way to judge the appropriateness of existing regulations and variances on these regulations for the scope of anticipated geological and operational conditions. The ability to make these objective determinations will rest on future research into the mechanical and geological problems involved, research leading to a body of knowledge that can be drawn upon when operating questions arise.

RESEARCH ON PREVENTION OF PIPELINE RUPTURE

Most of the ruptures of undersea pipelines result from corrosion of the external surface of the pipeline. The conditions leading to this corrosion vary widely, and the ability to predict corrosion rates requires extensive research on how water chemistry and bottom structure affect corrosion.

Pipeline inspection by divers is difficult, incomplete, and expensive. Research is needed to develop a device that from time to time can be put through a pipeline to measure the minimum thickness of the stress-bearing portion of the pipeline.

RESEARCH ON POLLUTION PREVENTION FROM TANKERS

Prevention of Oily Ballast Discharge

The load-on-top system of ballast water handling has reduced pollution from tankers on long voyages. As time goes on and older tankers are taken out of service, virtually all large tankers will employ the load-on-top system on long voyages. The dirty-ballast discharge problem will then exist only on tankers whose voyages are too short to allow use of the LOT procedure (and on ships using the LOT system during very rough voyages). It may be feasible to require tankers to remain loaded in port long enough to allow the oil and water to separate.

The development of oil-water separators capable of dealing with dirty ballast water at shoreside is also feasible. Some work has been done on separators (45), but much more is needed to achieve acceptable separation at the swift flow rates required.

The work done on development of segregated ballast tanks and membranes for keeping ballast water and oil separated has been quite limited. This research should be expedited so as to develop ships requiring a minimum amount of ballast water and very few segregated ballast tanks and to determine whether or not the membrane concept can be made operational.

Minimization of Tanker Casualties

Tanker Design. Work should be done to determine safe loading and operating procedures for any tanker under specified conditions. There has been much thought of late about reducing the pollution risk following tanker grounding by requiring that tankers use double bottoms without carrying cargo between the bottoms. Work also needs to be done to determine just how much this will reduce the pollution risk, and to determine on which routes it should be required.

During 1969 and 1970, there were an estimated 926 tanker groundings, collisions, and rammings (1). It is likely that many of these tanker casualties occurred because of the great stopping distance and low maneuverability of large tankers. Research on the stopping distance of tankers could be oriented towards the determination of propulsive arrangements that can stop the ships faster than existing conventional single-screw arrangements. Research on tanker maneuverability could lead to more effective control devices such as propellers that push sideways.

Navigational Aids. It can be expected that collision avoidance radar will be common to tankers in the near future and will reduce the number of collisions. Because of the potential benefits of satellite navigation, research and development of satellite navigation systems should be speeded up.

Although work is proceeding on harbor traffic control, the rate of progress has been quite slow. One reason for this is that operational methods for harbor traffic control are not very well known. Another is that the degree of ship control that should be exercised by a harbor traffic controller is unknown. Accelerated research in this area is essential.

RESEARCH ON OIL POLLUTION CLEANUP AND CONTROL TECHNOLOGY

Treating Agents and Associated Equipment

The two most promising treating agents for future research are sorbents and burning agents. This is not to say that dispersants are not important, but research on dispersants has reached a high level and seems to need no further emphasis.

Sorbent systems offer a possibility of overcoming the difficulty of containing oil in swift currents. The work done so far on sorbent systems has been oriented towards small spills. The development of sorbent systems for rather large spills in moderate to swift currents needs further study.

If an oil slick could be burned efficiently enough to prevent serious air pollution, the oil pollution problem would be greatly reduced. The prospects for burning large, uncontained slicks with burning agents are not promising. The possibility of burning a fairly thick slick held by a fireproof barrier is more interesting. This would require the development of fireproof barriers and of burning devices able to achieve complete and self-supporting combustion.

Some research has been done on the stimulation of biological degradation of oil by introducing micro-organisms into the natural environment or a possible waste oil facility. This possibility warrants further study.

Pollution Control Barriers

The developmental work on calm-water and high-seas oil-pollution control barriers has been extensive, yet the barriers developed to date cannot hold oil in strong currents; specialized designs are needed. As previously described, the essential difficulty with holding oil in strong currents is that of entrainment of oil bubbles in the water. With a simple barrier, it is unlikely that any barrier improvement will substantially reduce entrainment. There is, however, a possibility that a more complicated barrier can affect entrainment. For example, if barriers were placed both upstream and downstream from the oil slick, the shielding of the slick from the current by the upstream barrier might preclude or reduce entrainment loss. Research in this area is needed.

Oil Skimmers and Collectors

Although a substantial amount of research has been done to develop new skimmers and collectors, no full-scale sea trials of these devices have taken place in the presence of oil. Full-scale trials are needed to determine the effectiveness of the devices themselves and to gain more information on proper logistics for dealing with oil spills. These studies may lead to the conclusion that it is impractical to attempt to contain large slicks at sea, that the best way of dealing with them is to sweep through them with barrier-skimmer combinations. It can be anticipated that in some instances it will be best to collect the oil from a sweeping barrier, but that in others it will be more desirable to burn the oil if burners are available.

Tow Ships for Sweeping Operations

At the present time, the aids-to-navigation (A/N) buoys in U.S. waters are heavy steel buoys anchored to concrete blocks by heavy steel chains. The Coast Guard is responsible for keeping these buoys on station and operating. It services the buoys with large ships called buoy tenders. The USCG is currently studying the feasibility of replacing the A/N buoys with lighter buoys anchored with lightweight moorings and serviced by smaller, lighter boats. If this system is implemented, new buoy tenders will have to be built and stationed along the entire navigable coast of the United States. Since the USCG is one of the governmental agencies with oil pollution cleanup responsibility, it would be ideal if the lightweight buoy tenders were designed to tow oil-pollution control barriers as well. They would have to be capable of towing at low speed (about 0.5 knots) with steering control.

Bibliography

1. Porricelli, J.D., Keith, V.F., and Storch, R.L., "Tankers and the Ecology," Transactions of the Society of Naval Architects and Marine Engineers, Vol. 79, 1971, pp. 169-221.
2. Maritime Administration Tanker Construction Program, Volume 1, Draft Environmental Impact Statement, NTIS Report No. EIS 730392D.
3. Film on Pollution Control by New York City Fire Department, produced under Grant # 15080 FVP to the Marine Division, New York City Fire Department, from the U.S. Environmental Protection Agency.
4. The Georges Bank Petroleum Study, Vol. II, "Impact on New England Environmental Quality of Hypothetical Regional Petroleum Developments" by Offshore Oil Task Group, Massachusetts Institute of Technology, Report No. MITSG 73-5 February 1, 1973, pp. 58-59.
5. Lindenmuth, W.T., Miller, E.R., Hsu, C.C., "Studies of Oil Retention Boom Hydrodynamics," October 1970, Technical Report 7013-2.
6. Alpine Geophysical Associates, Inc., Project # 15080 FTU, Contract No. 14-12-860, May 1971, "Oil Pollution Incident Platform Charlie, Main Pass Block 41 Field, Louisiana."
7. Bleakley, W.B., OCS Orders 8 & 9. Source: *The Oil and Gas Journal*, August 21, 1972.
8. Raser, M.D. and Van Cleave, H.D., "Methods and Procedures for Preventing Oil Pollution from Onshore and Offshore Facilities," Proceedings of Joint Conference on Prevention and Control of Oil Spills, American Petroleum Institute, 1801 K St., N.W., Washington, D.C., 1971.
9. Warner, D.G., Exxon Company, U.S.A., "Spill Prevention in Offshore Petroleum Producing Facilities," Proceedings of Joint Conference on Prevention and Control of Oil Spills, March 13-15, 1973, Washington, D.C., American Petroleum Institute, Environmental Protection Agency, and United States Coast Guard.
10. "Oil Spillage–Santa Barbara, California," Hearing before the Subcommittee on Flood Control and Subcommittee on Rivers and Harbors of the

Committee on Public Works, House of Representatives, Ninety-First Congress, First Session, Santa Barbara, California, February 14, 1969.

11. Research Report, Pacific Northwest Laboratories, Battelle Memorial Institute, Richland, Washington, "Review of the Santa Barbara Channel Oil Pollution Incident," to Department of the Interior, Federal Water Pollution Control Administration and Department of Transportation, United States Coast Guard, Washington, D.C., July 18, 1969.
12. Turner, O.M., "Oil Spill Prevention Practices in Pipelines and Terminals," Proceedings of Joint Conference on Prevention and Control of Oil Spills, Washington, D.C., March 13-15, 1973. Sponsored by: American Petroleum Institute, Environmental Protection Agency, and United States Coast Guard, pp. 79-84.
13. Massachusetts Institute of Technology, Ocean Engineering Report No. 72-22, "The Isolation of Oil and Other Fluids in Tankers from Seawater Ballast Using Impermeable Membranes," Final Report, Vol. 1, December 1972.
14. Leonard, D.J., "Development of Tank Vessel Overfill Alarm Instruments," Proceedings of Joint Conference on Prevention and Control of Oil Spills, American Petroleum Institute, Washington, D.C., June 1971.
15. Letter from A. McKenzie to J.H. Milgram, April 6, 1973.
16. Notes of telephone conversation between J.H. Milgram and Captain Billy Smith, Gulf Oil Company, April 1973.
17. Notes of telephone conversation between J.H. Milgram and Mr. Arthur McKenzie, Senior Adviser, Marine Environmental Protection Division, Logistics Department, Exxon Corporation, April 13, 1973.
18. Cowan, E., *"Oil and Water," the Torrey Canyon Disaster*, J.B. Lippincott Company, Philadelphia and New York.
19. Smith, J.E., "Torrey Canyon," Pollution and Marine Life: A Report by the Plymouth Laboratory of the Marine Biological Association of the United Kingdom, Cambridge University Press.
20. Canevari, G.P., Esso Research & Engineering Company, "Development of the 'Next Generation' Chemical Dispersants," Proceedings of Joint Conference on Prevention and Control of Oil Spills, March 13-15, 1973, Washington, D.C., American Petroleum Institute, pp. 231-240.
21. James, W.P., "Environmental Aspects of a Supertanker Port on the Texas Gulf Coast," Texas A & M University, College Station, Texas, 1972.
22. Cablioch, L. "The Fight against Pollution by Oil on the Coast of Brittany," Proceedings of the Seminar on Water Pollution by Oil Way, 1970, Applied Science Publishers, Ltd., Ripple Road, Barking, Essex, England, 1973.
23. Bone, C. and Holmes, W., "Lessons from the Torrey Canyon," *New Scientist*, Vol. 39, 1968, p. 492.
24. Welson, Smith A., "The Problem of Oil Pollution of the Sea," *Advances in Marine Biology*, 1970, p. 215.
25. Crindell, A.M. and Traxler, R.W., "The Isolation and Characterization of Hydrocarbon–Utilizing Bacteria from Chedabucto Bay, Nova Scotia," Proceedings of Joint Conference on Prevention and Control of Oil Spills, American Petroleum Institute, Washington, D.C., March 1973.
26. Schatzberg, P. and Nagy, K.V., "Sorbents for Oil Spill Removal," Proceedings of Joint Conference on Prevention and Control of Oil Spills, Washington, D.C., June 15-17, 1971.

27. Schatzberg, P., "Investigation of Sorbents for Removing Oil Spills from Waters," USCG Report No. 724110.1/2/1.
28. Oxenham, J.P., Cochran, R.A., Hemphill, D.P., Scott, P.R. and Fraser, J.P., "Development of a Polyurethane Foam Marine Oil Recovery System," Proceedings of Joint Conference on Prevention and Control of Oil Spills, Washington, D.C., March 13-15, 1973.
29. Miller, E.R., Jr., "Development and Preliminary Design of a Sorbent Oil Recovery System," Proceedings of Joint Conference on Prevention and Control of Oil Spills, Washington, D.C., March 13-15, 1973.
30. Dorrler, J.S., "Sorbent System Development for Oil Spill Cleanup," Proceedings of Joint Conference on Prevention and Control of Oil Spills, Washington, D.C., March 13-15, 1973.
31. Newman, P.E. and Macbeth, N.I., "The Use of Booms as Barriers to Oil Pollution in Tidal Estuaries and Sheltered Waters," Seminar on Water Pollution by Oil, Applied Science Publishers, Ltd., Ripple Road, Barking, Essex, England, May, 1970.
32. Hoult, D.P., Cross, R.H., Milgram, J.H., Pollak, E.G. and Reynolds, H.J., "Concept Development of a Prototype Lightweight Oil Containment System for Use on the High Seas," USCG Report No. 714102/A/003.
33. Pordes, O. and Jongbloed, L.J.S., "Laboratory Investigations into the Sinking of Oil Spills with Particulate Solids," Proceedings of Joint Conference on Prevention and Control of Oil Spills, Washington, D.C., June 15-17, 1971.
34. Miller, E., Lindenmuth, W.T. and Altman, R., "Analysis of Lightweight Oil Containment System Sea Trials," Hydronautics, Inc., Technical Report 7220-1.
35. Reynolds, H.J., Pollak, E.G. and Milgram, J.H., Phase 2 Development of a Prototype Lightweight Oil Containment System. In Preparation.
36. Glaeser, J.L., "An Effective Oil Spill Containment–Recovery System for High Seas Use," Proceedings of Joint Conference on Prevention and Control of Oil Spills, Washington, D.C., March 13-15, 1973.
37. Hoult, D.P., "Containment of Oil Spills by Physical and Air Barriers," American Institute for Chemical Engineers Meeting, Puerto Rico, May 1970.
38. March, F.A. and Beach, R.L., "High Seas Oil Recovery System," Proceedings of Joint Conference on Prevention and Control of Oil Spills, Washington, D.C., March 13-15, 1973.
39. Leigh, J.T., LCDR, "Oil Recovery on the High Seas," Proceedings of Joint Conference on Prevention and Control of Oil Spills, Washington, D.C., March 13-15, 1973.
40. Fay, J.A., "Physical Processes in the Spread of Oil on a Water Surface," Proceedings of Joint Conference on Prevention and Control of Oil Spills, Washington, D.C., June 15-17, 1971.
41. Hoult, D.P., "Oil Spreading on the Sea," Annual Review of Fluid Mechanics, Vol. 4, 1972, Annual Reviews, Inc., Palo Alto, California.
42. Milgram, J.H., "Forces and Motions of a Flexible Floating Barrier," *Journal of Hydronautics*, Vol. 5, No. 2, April 1971.
43. Abrahams, R.N. and Miller, E.R., "Oil Spill Containment System Develop-

ment and Testing Program," Proceedings of Joint Conference on Prevention and Control of Oil Spills, Washington, D.C., March 13-15, 1973.

44. Arnold, K.E., "A Systems Approach to Offshore Facility Design," Third Annual Meeting, Division of Production, American Petroleum Institute, April 1973.
45. Harvey, A.C. and Stokes, V.K., "Evaluation of a Unique Centrifuge for Separation of Oil from Ship Discharge Water," Proceedings of Joint Conference on Prevention and Control of Oil Spills, Washington, D.C., March 13-15, 1973.
46. Wayland, R.G., "Federal Regulation of the United States Offshore Oil Industry," Proceedings of the Conference Sponsored by the International Association for Pollution Control, Washington, D.C., May 1972.
47. Exxon Corporation, "Preventing Oil Spills." (In press)
48. Nelson, R.F., "The Bay Marchand Fire," Shell Oil Company, Offshore Exploration and Production Division, October 4, 1971.
49. Freiberger, Arnold, "Burning Agents for Oil Spill Cleanup," Proceedings of Joint Conference on Prevention and Control of Oil Spills, Washington, D.C., June 15-17, 1971.
50. Goldstein, A.M., Koros, R.M. and Tarmy, B.L., "Engineering Study of an Oil Gellation Technique to Control Spills from Distressed Tankers," Proceedings of Joint Conference on Prevention and Control of Oil Spills, Washington, D.C., March 13-15, 1973.
51. Vedder, J.G., Wagner, H.C., Schoellhamer, J.E., Yerkes, R.F., Yenne, K.A., McCulloh, T.H., Hamilton, R.M., Brown, R.D., Jr., Burford, R.O., and DeNoyer, J.N., "Geology, Petroleum Development, Seismicity of the Santa Barbara Channel Region, California," geological survey, professional paper 679, United States Government Printing Office, Washington, D.C., 1969.
52. Wicks, M., "Fluid Dynamics of Floating Oil Containment by Mechanical Barriers in the Presence of Water Currents," Proceedings of Joint Conference on Prevention and Control of Oil Spills, American Petroleum Institute, 1800 K St., Washington, D.C., 1969.
53. Cross, R.H., and Hoult, D.P., "Collection of Oil Slicks," ASCE National Meeting on Transportation Engineering, Boston, Mass., July 1970.
54. U.S. Department of the Interior, Draft Environmental Impact Statement on Deepwater Ports, June 1973.
55. Crane, C. Lincoln, Jr., "Maneuvering Safety of Large Tankers: Stopping, Turning and Speed Selection," Transactions of the Society of Naval Architects and Marine Engineers, 1973.
56. Correspondence to the author from L.P. Haxby, Manager, Environmental Affairs, Shell Oil Company, May 1, 1973.

Index

About the Authors

Donald F. Boesch is a marine biologist at the Virginia Institute of Marine Science of the University of Virginia and a member of the Department of Marine Science at both the University of Virginia and the College of William and Mary. He was a Federal Water Pollution Agency Predoctoral Fellow in water pollution research and a Fulbright-Hays Postdoctoral Fellow in marine ecology at the University of Queensland, Brisbane, Australia. His research interests include marine ecology, particularly benthic ecology, and pollution ecology.

Carl H. Hershner is a Ph.D. candidate at the University of Virginia in marine biology. His research involves the study of chronic oil pollution of salt marshes in the Chesapeake Bay.

Jerome H. Milgram is a member of the Department of Ocean Engineering of MIT and specializes in the hydrodynamic aspects of ocean engineering, particularly those concerned with water waves. He is the designer of the U.S. Coast Guard High-Seas Lightweight Oil Pollution Containment Barrier and has had a major role in the at-sea testing of the barrier. He has also been responsible for measurements of the barrier's performance and the sea state in the tests of the barrier and oil recovery system. He has written extensively on the hydrodynamics of pollution control and the aerodynamics of sails.